A MOST VALUABLE MEDIUM

A MOST VALUABLE MEDIUM

The Remediation of Oral Performance on Early Commercial Recordings

—⁂—

RICHARD BAUMAN
WITH PATRICK FEASTER

INDIANA UNIVERSITY PRESS

This book is a publication of

Indiana University Press
Office of Scholarly Publishing
Herman B Wells Library 350
1320 East 10th Street
Bloomington, Indiana 47405 USA

iupress.org

First printing 2023

Cataloging information is available from the Library of Congress.

ISBN 978-0-253-06517-9 (cloth)
ISBN 978-0-253-06518-6 (paperback)
ISBN 978-0-253-06519-3 (e-book)

CONTENTS

ACKNOWLEDGMENTS

I OWE THANKS TO MANY colleagues for their generous contributions to the development of this book. My greatest debt is to Patrick Feaster, who first introduced me to the world of early commercial sound recording. Without Patrick's encyclopedic knowledge, his own extensive research, his great personal collection of early records, his technical know-how, and his unfailing generosity in guiding me through these materials, this book would not have been possible. I am especially grateful for Patrick's effort in assembling the sound files and discography to accompany the book and his discographic essay unpacking the complexities of early commercial record production.

Over the years that I have been engaged with early recordings, I have had numerous opportunities to present my work in progress in lecture form. For thoughtful and provocative comments on lecture presentations of the research that developed into the chapters that follow, I would like to thank Pertti Anttonen, Mary Bucholtz, Vânia Cardoso, Steve Coleman, Robson Corrêa de Camargo, Melissa Curtin, Michael Curtin, John Dawsey, Sandro Duranti, John Fenn, Jim Fox, Frog, Lisa Gilman, Jean Langdon, David Novak, Elinor Ochs, Joel Sherzer, Lotte Tarkka, and Mike Wilson. I owe a special debt of gratitude to Greg Urban, an anonymous reader for Indiana University Press, and my *compañero de trabajo*, Charles Briggs, for their insightful and productive comments on earlier drafts of the manuscript. Thanks, too, to Andy Kolovos, Steve Green, and Rachael Stoeltje for help in securing materials that proved important to the development of my thinking about Charles Ross Taggart. I am grateful to Don Brenneis for early guidance in sound studies; to Ilana Gershon for stimulating conversations about media anthropology; to Laurie Graham and T. M.

Scruggs for their gracious hospitality in Iowa City; and to Jason Jackson for his encouragement in moving the project along toward a book.

Throughout the gestation of this book and in so many other ways, my colleagues in the Michicagoan Faculty Seminar in Linguistic Anthropology have provided the best kind of intellectual community a scholar could wish for. Heartfelt thanks to Summerson Carr, Susan Gal, Matt Hull, Judy Irvine, Webb Keane, Alaina Lemon, Michael Lempert, Bruce Mannheim, Barb Meek, Costas Nakassis, Susan Philips, Michael Silverstein, and Kristina Wirtz.

And most of all, my deepest gratitude to my wife, colleague, sounding board, and dance partner, Beverly Stoeltje, for her unfailing support, even through the countless playings and replayings of these old recordings that she has had to endure. BJ, you are the best.

The author and publishers wish to thank the following for permission to use copyright material:

Chapter 1 is a revised version of "'Fellow Townsmen and My Noble Constituents!': Representations of Oratory on Early Commercial Recordings," with Patrick Feaster, *Oral Tradition* 20, no. 1 (2005): 35–57.

Chapter 2 is a revised version of "'Accordin' to the Gospel of Etymology': Burlesque Sermons on Early Commercial Sound Recordings," *Signs and Society* 6, no. 1 (2018): 166–204. © Semiosis Research Center, Hankuk University of Foreign Studies.

Chapter 3 is a revised version of "The Remediation of Storytelling: Narrative Performance on Early Commercial Sound Recordings," in *Telling Stories: Language, Narrative, and Social Life,* ed. Deborah Schiffrin, Anna De Fina, and Anastasia Nylund (Washington, DC: Georgetown University Press, 2010), 23–43.

Chapter 4 is a revised version of "The 'Talking Machine Storyteller': Cal Stewart and the Remediation of Storytelling," in *The Individual and Tradition: Folkloristic Perspectives,* ed. Ray Cashman, Tom Mould, and Pravina Shukla (Bloomington, IN: Folklore Institute, 2011), 71–91.

NOTE ON TRANSCRIPTION

IN TRANSCRIBING THE RECORDED PERFORMANCES analyzed in the chapters that follow, I have had two principal concerns in mind:

1. I intend the transcripts to convey that they are representations of spoken language. The chief means I have employed to this end is nonstandard spelling to capture features of pronunciation. I have not, however, resorted to eye dialect. One of the recurrent problems in transcribing oral speech—especially oral speech in nonstandard, vernacular dialects—is the danger of making the speakers appear to be unsophisticated rubes. I should make explicit, then, what will be even more obvious in the book: that those stereotypes are precisely what the *performers* are trying to convey, and if my transcriptions evoke them yet again, so much the better.

2. I have endeavored to represent by graphological means some of the significant formal patterning principles that organize the performances, chiefly prosodic structures but occasionally also episodic structures for certain narratives. Line breaks mark breath units, intonational units, or syntactic structures, which are usually—though not always—mutually aligned. Indented lines mark shorter pauses. Double spaces mark episode breaks.

LISTEN TO THE RECORDS

THIS BOOK IS AN EXERCISE in intermediality: it relies on the visual medium of print to analyze the acoustic medium of sound recording. Our hope is that the communicative affordances of print will allow us to provide illuminating insights into the world of early commercial sound recordings, but we are firmly convinced that it will be more illuminating still to offer you both, the visual and the acoustic. Accordingly, through the good offices of the Avalon Media System, Patrick Feaster has assembled an online repository of the recordings discussed in the chapters that follow, and we urge readers to make full use of it: listen to the recordings as you read about them. Links to the audio files of the recordings discussed in the book are provided in the body of the text and in the discography appended to the work. The entire collection is available at https://purl.dlib.indiana.edu/iudl/media/w62f26fr31.

A MOST VALUABLE MEDIUM

INTRODUCTION

"A Most Valuable Medium"

IN 1894, AT A TIME when he was preoccupied with raising capital for his fledgling United States Gramophone Company, the recording pioneer Emile Berliner advertised a promotional scheme that promised to boost the sales of his talking machine and yield a bit more immediate revenue as well, "The Gramophone as an Advertising Medium":

> Parties desiring to advertise their wares will find in the gramophone a most valuable medium.
>
> We will make for you any special plate, containing, besides an interesting musical piece, etc., a bit of advertising such as you may suggest; manufacture as many hard rubber copies as you may order at regular wholesale rates; and distribute them gratis to people buying Gramophones.
>
> ....
>
> Nobody will refuse to listen to a fine song or concert piece, or an oration— even if interrupted by a modest remark: **"Tartar's Baking Powder is the Best,"** or **"Wash the Baby with Orange Soap,"** etc. (Fabrizio and Paul 2002, 8; boldface in the original)

There are two commodities in play in Berliner's proposed entrepreneurial arrangement. The recordings, featuring engaging performance pieces and subsidized by the advertisers who sponsored them, would constitute a free premium to people buying Berliner's gramophone while they would promote the sponsors' products, such as baking powder or soap, touted in the brief commercial message appended to the performance.

An infusion of capital from investors in 1895 freed Berliner from having to pursue such small-scale fundraising schemes and allowed the reorganized Berliner Gramophone Company to concentrate on the business of manufacturing gramophones and producing records for home entertainment, a market sphere still in its nascent stages of development. Berliner was especially energetic in seeking out a variety of performance forms for presentation on his commercial recordings and was fortunate to engage as a talent scout the young Fred Gaisberg, later to become one of the most important figures in the development of the industry (J. N. Moore 1999). An early Gaisberg recruit to the Berliner roster of performers was the versatile, multitalented man of words George Graham. Interestingly, in an echo of Berliner's earlier advertising scheme, one of Graham's early recordings for the company was "Advertising Plant's Baking Powder" (Berliner 641), issued in 1896.[1]

As it happens, there was no Plant's Baking Powder on the commercial market. Graham's recorded performance was a simulation, a display of the pitchman's verbal skill, oriented not toward promoting an independent product but toward entertainment (Feaster 2007, 493). The advertising pitch itself—as an engaging, virtuosic performance—was the product, commoditized in the form of the record disk and available to consumers for fifty cents, on a par with all other Berliner commercial recordings for home entertainment. And indeed, the verbal artistry of pitchmen, auctioneers, carnival barkers, street vendors, and various other kinds of virtuosic sales agents—fluent, insistent, witty, hyperbolic, replete with parallel structures that catch up the hearer in patterns of expectation and fulfillment—proved to be an attractive resource for Berliner and other record producers. Remediated, repurposed, and recontextualized from an oral performance in the service of advertising a product to prospective customers on a street corner or marketplace to a mechanically mediated display form, commoditized for its entertainment value to customers in the comfort of their homes, the sales pitch as a performance genre (see Bauman 2004, 58–81) was featured on a number of early commercial recordings.

George Graham was, by all indications, a practiced pitchman. Gaisberg claims to have discovered him in that capacity, describing him as a raffish "character" who

> steered the easiest course through life, sometimes as a member of an Indian Medicine Troupe doing one-night stands in the spring and summer and in the winter selling quack medicines on the street corners. His tall, lanky figure, draped in a threadbare Prince Albert coat and adorned with a flowing tie, his wide-brimmed Stetson hat and his ready stream of wit combined to extract the dimes and nickels from his simple audience in exchange for a bottle of colored water. I discovered him one day on the corner of Seventh and Pennsylvania Avenue selling a liver cure to a crowd of spellbound negroes. He was assisted by John O'Terrell, who strummed the banjo and sang songs to draw the crowd. (Gaisberg 1942, 11)

Fig. 0.1. Patent medicine salesman. Photo by Marion Scott Wolcott for the Farm Security Administration, 1939, Library of Congress Prints and Photographs Division, Washington, DC.

Although Gaisberg's veracity is occasionally called into question (see Feaster 2007, 494), his description of Graham as pitchman rings true: the colorful dress, the street-corner venue, the "ready stream of wit" to hold the audience spellbound, the use of unrelated performance forms to draw a crowd, and so on.

During a series of early visits to the Berliner studios in the spring of 1896, Graham displayed his skills as a pitchman ("fakir" was the nineteenth-century colloquial term) on three recordings: "Street Fakir" (Berliner 638Y; https://purl.dlib.indiana.edu/iudl/media/r46q97fk2o) and "Fakir Selling Corn Cure" (Berliner 639; https://purl.dlib.indiana.edu/iudl/media/g35445jz5m), recorded on May 23, and "Advertising Plant's Baking Powder" (Berliner 641; https://purl.dlib.indiana.edu/iudl/media/c77s65px6z), recorded on May 26. These three recordings, produced in the brief span of four days, compose, in effect, an experimental set, in which Graham explores a variety of formats for the representation of sales pitches in a medium that relied solely on sound.

Consider first "Advertising Plant's Baking Powder," the recording I mentioned a moment ago. Here is the opening section of the recorded sales pitch:

**George Graham, "Advertising Plant's Baking Powder" (Berliner 641),
recorded May 26, 1895**

Now there, friends,
a few words to you about baking powder.
I wish to say something about Dr. Plant's Cream of Tartar Baking Powder.
This is absolutely the purest and best baking powder on the market today.
Now, the public are cautioned against buying these cheap powders
which contain alum and ammonia.
Now, you all know that ammonia is made from the hoofs of dead and decayed
   animals.
And friends, think of that!
Think of that!
Think of that!
A preparation like that being used in a baking powder.
Now, this baking powder,
Dr. Plant's Baking Powder,
is a pure cream of tartar baking powder.
Now, you want something to raise the bread properly,
you want good bread,
bread is the staff of life.
Yes.

As I suggested earlier, this performance is a simulation, a demonstration of the pitchman's art. My claim is based in part on the historical fact that there was no commercially available product called Plant's (or Dr. Plant's) Baking Powder, so there was no way that the recorded pitch could lead to a purchase of the product being advertised. Still, we can't assume that all listeners to the recording would have known that Plant's Baking Powder did not exist. For those who were un-aware that there was no such product, the recording might have appeared to be a genuine advertising pitch targeted at them. That is, after all, what the title would suggest, and Graham's performance displays characteristic thematic and stylistic features of the real thing. My suggestion that the performance was a simulation is also based in part on the fact that the recording was sold for fifty cents, the same as all of Berliner's commercial recordings for home entertainment, which would place it on a par with recordings of other, more obvious performance forms. We should note, however, that the conventional street-corner sales pitch was multifunctional. It served as a genuine advertise-ment for a real product, but it was also entertaining in its virtuosity and wit. Traditional pitchmen maintained that people often bought their products as a token of appreciation for the entertainment the performance afforded them

(Dargan and Zeitlin 1983, 24). From this standpoint, record buyers might take the recorded pitch as a true advertisement and still be willing to pay for the recording because of its entertainment value. There is, thus, a measure of ambiguity surrounding Graham's sales pitch for Plant's Baking Powder: Would listeners hear it as a genuine advertisement for a real product or as a simulation?

If we turn to another performance of the pitchman's verbal art recorded by Graham during the same four-day period as "Advertising Plant's Baking Powder," we find none of the framing ambiguity we encountered in that recording. Here is how "Street Fakir" starts off:

George Graham. "Street Fakir" (Berliner 638Y), recorded May 23, 1895

Imitation of a street fakir,
by George Graham.
Now, friends,
if you'll gather round and give me your kind and undivided attention for a few
    moments
I will endeavor to entertain and amuse you
by performance of several feats of legerdemain
commonly known as magic.
Now, a good many people seeing me appear upon the public thoroughfare
imagine that I have got something to sell.
That, I assure you, is not the case.
I am here simply to advertise,
to advertise and introduce a preparation
that has a reputation extending from ocean to ocean.
I refer to Dr. Boccaccio's celebrated Egyptian Liniment,
one of the grandest preparations ever invented.
It cures coughs, colds, sore throats, rheumatism, and neuralgia,
in fact, all aches and pains.

The framing of this recording is clear from the very opening lines: "*Imitation of a street fakir, by George Graham.*" In this introduction, a kind of acoustic label characteristic of many early recordings, the performer makes explicit that what the listener is about to hear is an imitation and foregrounds the identity of the performer (Feaster 2001, 78–80). As Graham launches his imitation, he immediately opens the performance to a wider representation of the pitchman's art in its customary context. "Advertising Plant's Baking Powder" starts right off with a pitch for the product, with the presence of an audience of potential customers invoked indirectly by the address formula "Now, friends." "Street Fakir" begins the representation at an earlier point, with what pitchmen call, in the jargon of the trade, "building the tip," that is, gathering the audience itself (Dargan and Zeitlin 1983, 17). To aid them in building the tip, pitchmen frequently utilized other display forms to catch the attention of passersby—recall Gaisberg's account of Graham selling liver cure on a Washington street corner, aided by a banjo player to draw the crowd. Here, Graham portrays his street fakir drawing in an audience with the promise of magic tricks, also a common lure for building the tip. And before getting to the product itself, Graham fills out the mise-en-scène of his simulated enactment by evoking the physical setting of the sales pitch "upon the public thoroughfare."

Toward the end of the recording, after extolling the curative powers of Doctor Boccaccio's liniment, Graham draws the performance to a close with an enactment of what pitchmen call "turning the tip," that is, urging customers to buy and concluding the sale (Dargan and Zeitlin 1983, 24):

> Now, people,
> the regular price of this preparation
> is one dollar per bottle.
> But today, in order to advertise it,
> I shall pass it out at the phenomenally low price
> of twenty-five cents per bottle.
> Anyone wish a bottle?
> Yes, sir.
> Thank you, sir,
> and thank you, sir,
> and I guarantee you'll derive more benefits from this quarter
> than many a quarter you have invested.
> Thank you, sir.
> You too, sir.
> Thank you very kindly, sir.
> Well, as I see the blue-coated guardian of the peace
> coming around the corner,
> I will now leave for fields and pastures new.

The success of the pitch is apparent, as Graham evokes the fakir making a series of rapid sales, indexed by his apparent responses to customers handing over payment: "Yes, sir. Thank you, sir, and thank you, sir." In the third recording of the set, "Fakir Selling Corn Cure," Graham enhances this interactional portrayal still further by use of an additional device, taking on a different voice for the pitchman and the several members of the crowd with whom he engages. Finally, he closes "Street Fakir" by announcing a hasty departure, as he notices a policeman coming around the corner, evoking yet again the street-corner setting as he takes his leave of it. What we have, then, in "Street Fakir" and "Fakir Selling Corn Cure" is an elementary form of audio theater, in which Graham manages to evoke physical setting, mise-en-scène, dramatis personae, actions, and interactions solely by sound, the acoustic medium of his voice and the evocative resonance of his talk. In "Advertising Plant's Baking Powder," the listener is ambiguously positioned between direct target of a sales pitch, picked out by "Now, friends," and audience member, watching—or listening to—a simulated enactment of a sales pitch. In "Street Fakir" and "Fakir Selling Corn Cure," the listener is straightforwardly a member of an audience taking in an artful representation of a street fakir at work.

This brief exploration of Berliner's and Graham's experimentation with ways of rendering the art of the pitchman on commercial recordings suggests the kinds of problems that motivate this book. Considered a bit more analytically, the problems that confronted the makers of these recordings had to do with the process of adapting a primarily oral and embodied performance form that participants were accustomed to engaging with in co-present situations to a new communicative medium that was restricted to sound alone and in which performer and audience were removed from each other in time and space (cf. Samuels et al. 2010, 337). This process of adaptation carried with it a number of additional formal and functional considerations. Sales pitches in their originary contexts were open-ended in duration, and virtuosic pitchmen could draw them out to considerable and impressive length (Bauman 2004, 67–74; 2012, 102–109), while commercial recordings in the early days of the industry were limited by the capacities of the technology to roughly two and a half to three minutes in length. Moreover, the performances of sales pitches oriented toward selling things took place in public space, on the street, in the market, designed to attract and hold the attention of potential customers, draw them in, and persuade them to buy. The recorded performances, by contrast, were actualized in private, domestic space for the purpose of home entertainment. And finally, at least for the moment, sales pitch performances in the street or the market were ephemeral; they were gone in the moment of enactment. On record, however, they became durable, preserved in material, collectible form, repeatable in new ways.

To be sure, these practical considerations of production and consumption are not at all limited to adapting the performance of sales pitches to commercial sound recordings. Indeed, I draw on this small, experimental set of early recordings chiefly because they open up to us in a clear and economical way the problems that early producers and performers of spoken-word recordings faced in adapting the full range of available verbal performance genres to the new technology of sound recording for a commercial market. In the chapters that follow, I focus on the three verbal genres that were arguably the most important in the late nineteenth- and early twentieth-century oral performance repertoire—oratory, sermons, and stories—taking as my point of departure the kinds of problems attendant upon the adaptation of these forms to a new medium and following them outward as they lead in a variety of social and cultural directions.

My approach in this work is closely aligned with what Jason Camlot (2019, 5) terms "phonopoetics," directed toward elucidating "the emergence and making (poesis) of literary speech sounds (phono) as they can be heard in early spoken recordings." Camlot's primary unit of analysis, like mine, is the recorded text (23). He argues for "a methodology of audiotextual criticism that does its utmost in the first instance to account for the relationship between media format and modified conceptions of genre or form as applied to the audiotext" (169). Importantly, beyond these formal and pragmatic considerations, Camlot's perspective, again like mine, extends to "the location of a sound recording within a specific historical context" (5). Where Camlot's primary focus is on the recording of literary works composed initially to be read, however, my interest centers on primarily oral genres and on oral poetics, examining the ways in which oral performances are made and how the process of making on the part of the performer is adapted to commercial sound recording.

In broader terms, I would identify this project as a study in *remediation*. I adopt this rubric, first developed systematically by Bolter and Grusin (1999), because it captures, in the prefix *re-*, the sequential relationship involved in the adaptation of an antecedent performance form actualized by means of a specific medium, the embodied semiosis of voice and gesture, to a new medium, recorded sound. Methodologically, however, my approach hews more closely to a related perspective on the interactions among cultural forms realized across media—namely, *intermediality* (Elleström 2010; Jensen 2016)—which opens the way to the closer focus on formal and pragmatic relationships and contrasts that I try to provide than on sequentiality. More specifically, I conceive of the process of remediation as a practical problem: How do performers adapt a known and familiar performance form for which there are well-established patterns of actualization to a new medium that calls for new modes of performative production, reception, and participatory engagement? What aspects of the former process carry over, and what changes?

To a degree, these questions arise in any act of recontextualization—lifting a text out of its originary discursive and situational context and recontextualizing it in another (Bauman 2004, 4–11; Bauman and Briggs 1992). When it comes to remediation, however, a new consideration comes forward—that is, the affordances of the new medium, the technical and communicative capacities and limiting factors inherent in the new medium as engaged by participants (Hutchby 2001, 30). The newness of a new medium, of course, is a historical phenomenon, anchored in time and place (Gershon 2017; Gershon and Bell 2013; Gershon and Manning 2014). For my purposes in this book, there were two relevant phases of phonographic newness. First, in the months that followed Edison's announcement of his marvelous new invention, observers, commentators, and even Edison himself could only speculate on what it might be used for. In several of the chapters to follow, I draw on Edison's early speculations as points of departure.

The threshold of newness that is more fully and immediately relevant to my purposes, however, came later, beginning in 1889, marked by a radical reorientation on the part of the recording industry from a preoccupation with the marketing of recording technology primarily for office use—essentially as a dictation machine for business and legal purposes—to the cultivation of sound recording as a medium of entertainment (Gelatt 1977, 45–57; Gitelman 2006, 44–57; Millard 1995, 42–49; Welch and Burt 1994, 87–95). The breakthrough factor was the realization on the part of an initially small but rapidly growing number of entrepreneurs that record playback machines could be coupled with coin-in-the-slot mechanisms adapted from weighing and vending machines to make recorded musical and verbal performances accessible to audiences listening through rubber ear tubes. Nickel-in-the-slot machines, placed in public spaces like drugstores, hotel lobbies, train and ferry terminals, saloons, and—before long—dedicated phonograph parlors, proved to be enormously popular and, still better, profitable. The popularity and profitability of these machines provided a stimulus for further orders of innovation, performative and technological, that paved the way for the next phase of development of sound recording as an entertainment medium.

Nickel-in-the-slot record offerings included a range of performance forms, especially comic songs, band music, instrumental solos, and humorous monologues and recitations (Walsh and Burt 1994, 90). The latter two forms, already part of early vaudeville repertoires, constitute the focus of this book. The period between 1889 and the mid-1890s, as the entertainment potential of sound recordings came fully into view, emerged as a preliminary proving ground for performers who became, from 1895 onward, core members of the corps of recording artists featured on commercial recordings for home entertainment, figures like Russell Hunting, Len Spencer, and Dan Kelly. It was on recordings made for nickel-in-the-slot machines that performers who were attuned early to the potentials of the new medium began to adapt their performances to the affordances of the technology—to adjust to the time limits imposed by the carrying capacity of early recording formats, to modulate the register and timbre of their voices, to monitor their methods of enunciation and their speech rate, and so on—in ways that were best suited to the capacities of sound recording technology. Thus, the modes of performance that characterize the recordings produced from 1895 onward for a domestic market were, to a degree, preadapted to the nascent mass medium of home entertainment. Perhaps more important for the commercial development of sound recordings for "the home circle" (Gelatt 1977, 85), though less so for my interests in this book, was duplicating capacity, the technology that would allow for the production of multiple copies of a recording on a scale sufficient to serve a mass market (Gelatt 1977, 81, 87–89; Millard 1995, 44–49). Efforts to develop this technological capacity accelerated throughout the early 1890s and reached the point of commercial viability around 1895, paving the way for the full realization of sound recording as a mass medium of home entertainment. And finally, the last key factor in getting recorded performances into homes was the production of affordable playback machines. Here too, the watershed developments occurred in the latter part of the 1890s (Gelatt 1977, 70–71). From introductory price levels of $40–$50 in 1895–1896, the cost of basic machines to serve as "an entertainer in the home" (Fabrizio and Paul 2002, 11) fell rapidly to $20–$25 by 1897, until, by 1899, consumers could buy a serviceable Edison phonograph for $7.50. Edison's pronouncement in 1907 that "there is no family so poor that it cannot buy a talking machine" (quoted in Millard 1995, 54) was hyperbolic, to be sure, but it underscored the economic basis necessary to secure the place of the phonograph as a mass medium of home entertainment.

As I suggested at the beginning of this introduction, the period of exploration, discovery, and experimentation for participants in the remediation of performance from co-present and embodied to commercial sound recording coalesced in the mid-1890s and extended to around 1920. The technological and institutional history of the nascent commercial record industry has been well documented, and I won't focus on it here (see Chanan 1995; Gelatt 1997; Gitelman 1999; 2006; Kenney 1999; Millard 1995; J. N. Moore 1999; Morton 2000; Sterne 2003; Suisman 2009; Welch and Burt 1994). For the purposes of this book, the mid-1890s is the point at which we begin to have a sufficient corpus of surviving records to sustain broad examination of the emergent medium, while 1920 marks roughly the transition point between mechanical, acoustic recording technology and electronic recording, which changed the affordances of the medium at the same time that it marks the beginning of the era of broadcast radio, a new mass medium of home entertainment with transformative effects on the record industry.

In broader historical terms, 1895–1920 was a period of intense and widespread social change—economic, demographic, political—the dynamics of which inflected and interacted with the dynamics of remediation that represent my principal focus in this book (cf. Nakassis 2016, 231–237). Chapter 1, for example, explores the experimental efforts of political candidates and record companies, beginning in 1896, to tap the potential of sound recording as a means of expanding the reach of campaign speeches, ultimately opening the way to the commoditization of political engagement within the polity. The analytical focus of this exploration is the ways in which recorded campaign speeches are aligned to different orders of publics, as candidates and other participants in the political process experiment with various forms of addressivity potentiated by the new medium of sound recording.

Chapter 2 focuses attention on burlesque sermons drawn from the tradition of blackface minstrelsy. These parodic send-ups of African American preaching were poised between negative and painful stereotypes born of White racism and the culturally redemptive potential of traditional preaching for the creation of a new African American literature.

In chapter 3, the analytical focus shifts to the range of presentational formats employed in adapting traditional storytelling to commercial phonograph records, examining the formal and functional correlates of frameworks ranging from apparent direct narration to the listening audience to metanarrational accounts of storytelling events to audio theater. Storytelling remains the subject of chapter 4, which traces the performance career of the early recording star Cal Stewart, from his early days entertaining fellow railroad workers with his stories to his emergence as "the Talking Machine Storyteller," the first verbal performer to achieve stardom as a performer on commercial sound recordings. The final chapter, chapter 5, is devoted to Charles Ross Taggart, a popular performer on the lyceum and Chautauqua circuits who adapted his routines as the "Man from Vermont" and the "Old Country Fiddler" to the new medium of sound recording as a means of boosting his career as a platform performer. Taggart's recordings are especially noteworthy for their representations of rural ways of speaking and traditional fiddling as emblematic of fading ways of life as situated in time and space.

—m—

The study of remediation and allied concepts has a piecemeal history, approached from a variety of loosely coordinated directions and marked by a blinkered unevenness of substantive focus. The literature on remediation is devoted for the most part to digital media, and even in charting the broad historical scope of transformational processes by which extant communicative forms and practices are adapted to new communicative media, there is a curious tendency to foreground the advent of certain technologies of communication and pass over others. For example, in listing the "important media of the twentieth century," Jay David Bolter (2016, 1759), coauthor of the charter document in the field of remediation, includes radio, film, television, and print in various forms. Enumerating "the media known for much of the 20th century as 'mass media,'" Klaus Bruhn Jensen (2016, 6), a leading authority in intermediality studies, lists printed books and newspapers, film, radio, and television. These are typical, authoritative formulations. Notice what's missing? Neither of these overviews, or indeed, most others, can find a place for the phonograph, hailed at the time of its invention as a brilliant, transformative breakthrough in human communicative capacity, to become—in the last few years of the nineteenth century and the early decades of the twentieth—the first mass medium of home entertainment.[2] My point is not that the remediations attendant upon the advent of phonograph recording warrant attention simply to fill a gap in the scholarly record. Rather, the history of communication is a history of remediation all the way down, as established ways and means of communication are adapted to new communicative technologies, each with its own affordances and social relations of production, consumption, and circulation. All of those processes of adaptation require new forms of communicative work, exploring what may be carried over from antecedent forms in relation to what emergent forms and practices are forged to manage communication in new media. I am most interested here in how verbal performance forms, grounded in co-present contexts of interaction, were adapted to commercial sound recording, but those adaptations and resultant forms and practices themselves formed the ground on which radio—the next mass medium of home entertainment—took shape from the early 1920s onward. Nevertheless, as I demonstrate in chapter 3, performance features that were developed by early record performers are commonly attributed to the emergence of broadcast radio. As ethnographers and historians of media and performance, it behooves us to investigate each link in the remediational chain. This book, then, is an exploratory attempt to insert Edison's marvelous talking machine into the remediational record through close examination of how oral performance forms were adapted to this remarkable new medium.

NOTES

1. I draw closely on conversations with Patrick Feaster in the following discussion of Graham's recorded sales pitches. Feaster (2007, 494–501) contains a fuller and more thoroughly contextualized account of these materials.

2. For a similarly grounded critique of media historians' slighting of sound media, see J. Smith 2015, 2.

ONE

—w—

"COME IN *HERE* AND *HEAR* THEM SPEAK!"

Campaign Speeches and Political Publics

WITH PATRICK FEASTER

INTRODUCTION

When Thomas Edison listed the potential applications of his new invention to an eager public, far down the list, after talking dolls, other mechanical toys, and alarm clocks, was "Speech and Other Utterances" (Edison 1878, 534). It may have been unclear just what the commercial potential might be for recorded speech, but what Edison anticipated was the usefulness of the phonograph as a technology of commemoration: "It will henceforth be possible to preserve for future generations the voices as well as the words of our Washingtons, our Lincolns, our Gladstones, etc., and to have them give us their 'greatest effort' in every town and hamlet around the country, upon our holidays" (1878, 534). The preservation of great oratory and its reproduction on ceremonial occasions seemed an appropriate and desirable use for the phonograph. This was speech worthy of fixing and storing up not just as words—which could be accomplished in print—but as performance, in its living voice. Journalists, too, in their first excited accounts of Edison's "acoustic marvel," envisioned the preservation of great oratorical voices as a fitting use for the new technology. "How startling it will be," exclaimed an early article in *Scribner's Monthly*, "to reproduce and hear at pleasure the voice of the dead," including prominently "the speeches of celebrated orators" (Prescott 1877, 857).

In an article titled "The Phonograph" in the November 7, 1877, issue of the *New York Times*,[1] the use of the phonograph for the preservation and reproduction of oratory moves to the fore. Playfully developing the conceit that the phonograph "bottles up" speech for future use, the author suggests that "it may seem improbable that a hundred years hence people will be able to hear the voice of

13

WENDELL PHILLIPS in the act of delivering an oration, but the phonograph will render it possible to preserve for any length of time the words and tones of any orator."[2] To this author, "it is evident that this invention will lead to important changes in our social customs." The principal change, however playfully it may be framed, amounts to the recontextualization of public culture to private settings in commoditized form: "The lecturer will no longer require his audience to meet him in a public hall, but will sell his lectures in quart bottles, at fifty cents each; and the politician, instead of howling himself hoarse on the platform, will have a pint of his best speech put into the hands of each of his constituents." While the author of the article in *Scribner's Monthly*, like Edison himself, foresaw the use of the phonograph as among the "public uses" (Prescott 1877, 857) of the technology, in keeping with the public context of oratorical performance, the *Times* article anticipates the movement of public oratory to the domestic space, "the home circle." What follows logically, then, is the possibility that a private individual might build up a collection of recorded speeches containing a mixture of oratorical styles, as one develops a private wine cellar, with all the associated trappings of connoisseurship and consumerism. To speculate thus in terms of the "oratorical cellar" and the "connoisseur of orators" is to anticipate an affluent audience for sound recordings, those who could afford prestige goods made for the burgeoning consumer market. But it is not only the pleasure and pride of connoisseurship that that might follow from the "bottling" of oratory. The author points toward another significant potential of bottling speeches— that is, that a recording might stand in for a living speaker. All these innovations and potentials will figure in the explorations of recorded oratory to follow.

The recordings of political oratory at the center of this chapter fall into two major categories: (1) recitations of canonical speeches from American history and (2) campaign speeches by candidates for election to public office. We will be concerned in our examination of these materials with the transformations attendant on the process of representation, here including the effects and concomitants of mediation, the effects of semiotic reduction to sound alone, the recontextualization of oratory from public to domestic space, and the constraints imposed by the technological limitations of the medium. We are especially interested, though, in the rekeying and refiguration of participant structures and roles: How do the recorded performances align themselves to an audience? By audience here, we mean the targeted receivers of the performance (though not necessarily the addressees), invited to hold the performer's act of communicative display in close attention and to evaluate the skill and efficacy with which the performance is accomplished (Bauman 1977, 2012). And because the performances we are dealing with center on political oratory, how do the oratorical performances align themselves to a public (or to publics, in the plural)—both presupposed, in the sense of already recognized social forma-tions, and emergent, constituted by the recordings themselves and the market-ing efforts that promoted them?[3]

The term *public*, as we all know too well, covers a shifting and often in-choate field of phenomena, so in the interest of making explicit what we are about, let us specify also what we mean by the term. We take public in the nominal sense—a public—as a social formation constituted by discourse oriented to the life in common of a collectivity and constructed to foster dis-semination, either synchronically, through open accessibility and direction to multiple addressees, or diachronically, through expansive or accelerated circulation, or both (cf. Hénaff and Strong 2001, 1; Urban 2001). Different orders of metapragmatic regimentation will constitute different publics or constitute the same publics on different grounds. The regimenting factors may involve sites of discursive production, generic or textual form, addressiv-ity, and others, to be discovered in any empirical instance. All of these, of course, will be closely bound up with the capacities of the communicative technologies employed.

REANIMATIONS OF CANONICAL SPEECHES

The first category of recorded performances, recitations of famous speeches from the historical canon, consists of reanimations of the words of others, recontextualizations of the memorable utterances of famous orators, lifted out of their originary contexts of production and reperformed in new ones. The speeches continue to be attributed to their absent authors and associated with the occasions on which they were originally delivered, but in the guise in which we now hear them, on records, they are decoupled from both. The current re-citer is not accountable for the message, only for the delivery.

Let's consider a couple of examples. Lincoln's Gettysburg Address was re-corded regularly throughout the 1890s and early 1900s by early performance specialists in the new medium of sound recording, including W. F. Hooley (https://purl.dlib.indiana.edu/iudl/media/197x61mb69) and Len Spencer (https://purl.dlib.indiana.edu/iudl/media/g94h640p74) (Gracyk 2000, 180–183, 314–319, 328). The Gettysburg Address was a natural: not only was it the most widely known piece of American oratory in the repertoire, but it was short enough to fit in its entirety on a single record, which at the turn of the century meant two or three minutes.

From one point of view, we can recognize these recordings as belonging to an endless series of reiterations of this canonical speech. By the time com-mercial sound recording became a reality, the Gettysburg Address had been memorized and declaimed by generations of schoolchildren and students of elocution, performed in school exhibitions and other performance occasions. It is the quintessential commemorative text (Casey 2000, 216–257): as delivered by Lincoln in 1863, it commemorated the death of the battle victims and the birth of the nation four score and seven years earlier, and as reperformed by those generations of reciters, it commemorated Abraham Lincoln as well, the martyred hero who gave his life in the service of liberty and national unity. The Gettysburg Address thus represents the perduring ancestral word, recited on ceremonial occasions to commemorate the ancestors and available as well as a means of ceremonializing any occasion through intertextual ties with past ceremonies in which the speech was recited. Moreover, the Gettysburg Address is the authoritative word, as actively manifested in the verbatim replication of the text and the virtuosic crafting of the recitation, subject to evaluation for the relative skill and affecting power of the delivery (Bakhtin 1981, 342; Bauman 2004, 150–153).

Virtuosic performance displays high regard for the authoritative text, represents it as worthy of reproducing artfully, with care (Bauman 2004, 150–151). Spencer, Hunting, and other phonographic orators use a declamatory style promulgated by nineteenth-century elocutionists (N. Johnson 1993), marked by a slow, solemn pace; hyperprecise enunciation; careful marking of word boundaries; lengthened, resonant vowels (with an occasional quaver to signal affect); frequent use of tapped and trilled r's; measured intonation patterns; and so on. The style serves both as a vehicle for the display of artistry and as an index of solemnity.

How is this recording aligned toward a public? First of all, hearing the phonographic performance evokes those past ceremonial and performance occasions in which one has heard the Gettysburg Address before, as part of an assembled group of coparticipants in a public event, public in the sense of taking place in public space, openly accessible, on view, collectively enacted. Let's call this an *assembled* (Agacinski 2001, 137) or *gathered public*. Second, the phonograph's reiteration of the speech, and the recognition that it is a reiteration, invokes a *historically founded public*, made up of those who are heir to the legacy of the memorialized ancestors. And third, it invokes what we might call a *distributive public*, constituted by the dissemination of the text: those who have active or passive knowledge of it as a text and as a sign.

The siting of the recorded performance—that is, the playing of the record—in domestic space is of less transformative significance than one might assume. Many households of the period had print versions of the Gettysburg Address, in schoolbooks and anthologies, and, more important for our purposes, domestic declamations of the speech were common; it was an elocutionary display piece, and this was an era of elocutionary cultivation in the service of upward social mobility—as a tool for success in business and the professions—aided by teachers of elocution, self-help books, and other means (N. Johnson 1993). Thus, performances of the Gettysburg Address, Patrick Henry's "Give Me Liberty or Give Me Death," and other like pieces were brought into domestic space in mediated form well before the advent of sound recordings. So extensive was the distributive public of the Gettysburg Address that performers could rely on audience recognition of the source of parodic transformations, such as Byron G. Harlan's comic rendition in "Congressman Filkin's Home-Coming": "I come before you as a public servant who has worked for the people, by the people, and the people."[4]

A comparison of recordings of the Gettysburg Address with another oratorical staple of early record catalogs is revealing: "Portions of the Last Speech of President McKinley" (on Victor; https://purl.dlib.indiana.edu/iudl/media/m31168f386), also known as "President McKinley's Pan American Speech" (on Columbia). The speech was delivered at the Pan American Exposition in Buffalo, New York, on September 5, 1901, the day before McKinley was shot by the anarchist Leon Czolgosz. (He died on September 14.) Companies marketed recordings of parts of this speech within a few months of the assassination, but they continued to record new versions for at least another year or two and kept these in production for years thereafter. The McKinley selection stayed in the Columbia catalog until 1914 and in the Victor catalog until 1911 (Walsh 1971, 50, 92). Why continue to offer a recitation of portions of this speech so long after the fact? McKinley was an accomplished orator in his day, and it is likely that his reputation remained alive in the decade following his death. The proven long-term appeal of a recorded speech by one slain president—Lincoln's Gettysburg Address—may have encouraged the record companies to think in similar terms about a speech by a second presidential victim of an assassin's bullet. In any event, the recitation of McKinley's speech was, like the recitation of the Gettysburg Address, a commemorative act, exploiting the same authorizing and valorizing devices in performance and reaching back in time to an originary utterance.

It is noteworthy, however, that the recorded performance contains only about one-eighth of McKinley's original text (Hazeltine 1902, 10505–10512), which is all that could fit on a single recording. The portions selected for recitation turn out to focus on employment and trade conditions and their policy implications. Labor and tariff issues were central concerns of McKinley's political career, to be sure, but at the time our examples were pressed, around seven years after McKinley's death, the country was in a severe state of political instability that came to be known as the Panic of 1907, marked by economic failures, a depressed labor market, and trade anxiety. That is to say that in addition to their links with the ancestral past, the portions of McKinley's speech replicated on the recording invited recognition of the *current* salience of his message. What we are suggesting is that in addition to the alignment of this recording to the historically founded public that was heir to the legacy of McKinley's life and death, and to the distributive public constituted by the circulation of his last speech—both founded on the commemorative thrust of Spencer's recitation—this recording is aligned as well to a public constituted around an orientation to issues that bear upon their lives in common, perhaps the *polity as public.*

1896: BRYAN AND MCKINLEY

The earliest notice we have of commercial recordings of speeches keyed to a current political campaign comes from a catalog of the United States Phonograph Company issued during or shortly after the presidential campaign of 1896. The catalog notice of New Talking Records lists four speeches by William Jennings Bryan, the Democratic candidate, and one by William McKinley, the Republican nominee.

> HON. W. J. BRYAN'S CROWN OF THORNS AND CROSS OF GOLD SPEECH. The Peroration of the famous Address that won him the Presidential Nomination at Chicago. Very loud and distinct. Applause. No Announcement.

> MAJOR McKINLEY'S SPEECH ON THE THREAT TO DEBASE THE NATIONAL CURRENCY. As delivered by the distinguished Republican Nominee at Canton, July 11th. Very loud and distinct. Applause. No Announcement.

> HON. W. J. BRYAN'S SPEECH AT THE NOTIFICATION MEETING IN NEW YORK. A part of his Address at the great Demonstration in Madison Square Garden, New York, on August 12th. Very loud. Applause. No Announcement.

> HON. W. J. BRYAN'S REPLY TO THE CHARGE OF ANARCHY.

> From the Candidate's great Speech in Hornellsville, before 15,000 people in the open air. Very loud and distinct. Applause. No Announcement.

> HON. W. J. BRYAN'S OPINION OF THE WALL STREET GOLD-BUGS AND SYNDICATES. As delivered at the Buffalo Ratification Meeting, where he declared that the Creator did not make Financiers of better mud than he used for other people. Very loud and distinct. Applause. No Announcement.

The description of Bryan's "Cross of Gold" speech as "famous" in the first listing acknowledges the extensive publicity that the speech—and certainly the peroration—received from the very point of its delivery at the Chicago convention: "You shall not press down upon the brow of labor this crown of thorns;

you shall not crucify mankind upon a cross of gold." Further evidence of the speech's rapid achievement of a broad distributive public is the embedding of the peroration, quoted accurately, in a longer burlesque campaign speech by George Graham on a recording issued within months of the convention and less than one month after the election.[5]

The catalog listings, though brief, do a lot of contextualizing work, linking each recorded bit of oratory to one or another of the respective candidates, the larger speech from which it was taken, the event and site at which the speech was delivered, even its place in the ongoing campaign dialogue of charge and countercharge, as in "Hon. W. J. Bryan's Reply to the Charge of Anarchy." Note that each listing also contains the descriptive note: "Applause. No announcement." The latter point refers to the early convention of announcing the title, performer, and record company at the beginning of each recording; the departure of these recordings from the convention requires acknowledgment.

Interestingly, all of this contextualizing work serves to establish certain dimensions of ambiguity concerning the recordings. On the one hand, it allows for the interpretation that these recordings were made in situ, at the public events where the speeches were delivered, and feature the candidates themselves as speakers before a co-present audience. Note the framing of the McKinley recording that has been posted to YouTube:[6] "William McKinley: As the Republican candidate in the 1896 presidential election. Giving a [sic] 1896 campaign speech from his front porch." It points, in other words, toward what was still the default situational context for political oratory: a large-scale, heightened, formal (Irvine 1979) platform event (Goffman 1981, 165; 1983) involving a featured performer addressing a gathered audience, which expresses its appreciation of the message and the performance by means of applause.

But these recordings were, in fact, simulations, and, as the YouTube video suggests, they were convincing to record audiences, even to this day. They were reanimations of extracts from the candidates' speeches, most likely by Len Spencer, an early studio performer who worked for the United States Phonograph Company at the time. These factors point in the opposite direction, toward the detachability of texts from their originary settings and recontextualizability in other contexts. To be sure, this was not news: there was a long history, reaching back to classical antiquity, of inscribing speeches in writing, preserving them for their literary and historical interest, and reanimating them in recitation. What is interesting here is the element of simulation, the reenactment of the performance event allowed for by the capacity of the technology to reproduce the living human voice and by the inclusion of applause—an early form of mediated political simulacrum.

This capacity of the new technology of sound recording for dissimulation was recognized quite early, almost immediately, in fact, after the phonograph was invented, with political oratory as the object of fabrication. One of the first commercial exploitations of the phonograph was as popular entertainment: exhibitors took the new recording machines on tour, giving demonstrations before enthusiastic audiences eager to see—and hear—this marvelous invention. There is an account of one such exhibition in the *St. Louis Evening Post* of May 30, 1878:[7] "'Now, then,' said the gentleman in charge of the phonograph, to the crowd of spectators, 'we will have a mass-meeting.'" The speaker then turns to the machine:

> "Fellow citizens," begins the operator in a high key as if addressing a crowd of 10,000 people from the Court-house steps, "we have met here this evening to discuss the political situation, and as the first speaker who will address you I have the honor of introducing Hon. Berry Mitchell, of Cashokia Creek, who will address you on the issues of the day. Before the gentleman begins I propose three cheers for Mr. Mitchell, which I know you will give. Now, again, hip, hip, hurrah. Now once more to close up on."
> Into the ear of the phonograph the gentleman pours all these excited utterances. He then changes his talk. Assuming another voice, supposably from some disgruntled member in the crowd, he calls out, as people always do at political meetings, "Put him out." "Let's hang him." "Pull down his vest." "Down with the fraud."

The exhibitor then calls for music to calm the crowd, and a cornet player comes forward to play a strain from "Garry Owen." The exhibitor again steps forward "and indulges in a loud and ironical laugh, supposed to come from some scornful member of the crowd, who repudiates the speakers and the music, and despises in advance the political sentiments that are about to be promulgated." The recording is then played back: "'Fellow citizens, we have met here this evening,' the exact tone of the speaker being imitated perfectly, and then come the scornful remarks and the derisive laughter, the cheers, the hoots and yells, and all the usual accompaniments of a political meeting, including the music, which is reproduced perfectly." This is a remarkable performance, a simulated enactment of a political meeting featuring an oratorical performance. But it is also an illusion: one man, assisted by a musician, enacting multiple roles, contributing multiple voices to the recording. Moreover, the simulation is a highly condensed representation of a political meeting, employing a few diagnostic devices of the typical performance event and its constituent genres, which are so fully familiar to the audience that they are able instantly to recognize what is being enacted, aided, to be sure, by the performer's framing announcement, "we will have a mass-meeting." The introduction of the speaker, the call for three cheers, the heckling, the music—"all the usual accompaniments of a political meeting"—are indexical icons par excellence of the real thing.

Note, then, the contrastive yet complementary constructions of recorded oratory that coalesced almost from the moment of Edison's invention. On the one hand, a rhetoric of speech "faithfully," "accurately," "exactly" reproduced, reproduced with "fidelity" (e.g., Edison 1878, 530; E. Johnson 1877, 304; Prescott 1877, 848; Anon. 1878, 1828); on the other, a demonstration that lays bare the technology's capacity for simulation and illusion. We return to this point later in the chapter.

THE CAMPAIGN OF 1900

With the approach of the 1900 campaign, the nascent sound recording industry and political managers began to think in more imaginative ways about the potential applications of the new technology to political campaigns, perhaps even leading to "a complete revolution in campaigning methods," in the suggestion of an article in *Phonoscope*, a journal that served the fledgling recording business ("Talking Machines" 1900, 6). The article is somewhat tongue-in-cheek, but the imaginings to which it gives expression suggest the range of possibilities that the new technology might serve. "It is now suggested," reports the article, "that instead of making a laborious campaign, candidates devote their time at home talking into a funnel and leave the campaign committees and the Phonographs to distribute their views to an admiring public" (6). Playing precisely on the mediated quality of sound recording, the *Phonoscope* article suggests the advantage that might accrue from the decoupling of voice from co-present, embodied speakers, the mitigation of the risks that inevitably shadow both performers and their audiences (Bauman 2012, 102): "Timid aspirants to office can obviate the embarrassment of facing an audience of doubtful sympathy, while the audience run no risk of shock either from the appearance or mannerisms of the speaker."

The *Phonoscope* for April 1900 ("New Use" 1900, 7) reported, "The Republican National Committee have a plan under way now by which reproductions of political speeches will be made for the Graphophone and Phonograph, and they will figure largely in the present campaign." The article envisions that the strategy would not be confined to one party alone but that "orators of renown . . . will make records of their most famous efforts, and same will be distributed broadcast for the edification of the wavering voter" (8). Some Washington Democrats devised an organized project to distribute speeches by Democratic orators to Democratic organizations around the country, "thus affording small rural localities that would not be visited by great political lights the privilege of hearing the questions of the hour discussed by these national celebrities in their own voices, the same as though they were actually present" ("Democratic Bureau's Suit" 1900, 7). Apparently, they secured recordings of short speeches by various luminaries, including Bryan, their presidential candidate; Adlai E. Stevenson, their candidate for vice president; and endorsements from others, such as Senator Murphy of New York; J. G. Johnson, chair of the National Executive Democratic Committee; and William Randolph Hearst, newspaper publisher and president of the National Association of Democratic Clubs. The project seems to have foundered, though, for lack of sufficient funds to see it through. And McKinley, standing on the dignity of his office, decided that "it would be highly improper for him to talk into the machines" and so quashed the plans of those campaign visionaries who had been urging him to do so ("No Talking Machines" 1900, 8).

Although these campaign recording projects do not appear to have come to fruition, the bases and terms by which they were envisioned are revealing. A significant part of the medium's appeal lay in its anticipated multiplying effect: speeches by star orators might be reproduced and widely used, at "every cross road and corner grocery throughout the land" ("New Use" 1900, 8), with the recorded versions, reproduced in many copies, standing as surrogates for the political orators themselves. Here, it is the medium's capacity to exploit and multiply the power of presence by reproducing the candidates' "own voices" that represents its greatest attraction. The passage quoted above is a benchmark use of the term *broadcast* in reference to the capacity of sound media, a metaphor drawn from the agrarian sense of the term: sowing seeds by scattering them widely. It has become a dead metaphor for us, but in this early usage—indeed, the earliest we have found—it pointed up the expansive communicative potential of recorded sound to carve out broad, dispersed publics constituted by listening in common—though not together or necessarily at the same time—to the same speaker.

THE 1906 HEARST GUBERNATORIAL CAMPAIGN

We again encounter the recording of campaign speeches in the New York State gubernatorial election of 1906. William Randolph Hearst, ever the mass media innovator, was the candidate of the Independence League, running against the Republican candidate Charles Evans Hughes. In an idiom reminiscent of the "bottled" speech trope coined in its own pages forty years earlier, the *New York Times* of October 10, 1906, reports, "Hearst Speech 'Canned' for Up-State

Farmers. He Talks It and Gestures It into Phonograph and Camera. A 12-Cylinder Harangue. The Absent-Treatment Candidate Will Be Projected in Sound and Shadow before the Voters of the Remoter Regions" (NYT Oct. 10, 1906, 3). The article goes on to say, "A canned Hearst speech is the latest wrinkle in the up-State campaign of the Independence League's editor-candidate. Mr. Hearst will try it on hamlets and villages in remote sections of the State which he either will not have the time to visit or which his luxuriously appointed special train cannot reach for the reason that there is no railroad leading to them." Hearst, we recall, was one of the luminaries recorded in the abortive plan by the Democratic Party to circulate recorded campaign speeches in the election of 1900. Now, it appears, he put the plan into action in his own gubernatorial campaign.

Recognizing that the votes of upstate farmers, "born to the Republican Party," would be critical to his hopes of being elected, Hearst's campaign "conceived the idea of reaching the voters with talking machine records and moving pictures" (NYT Oct. 10, 1906, 1). Accordingly, Hearst "talked at the graphophone against the trusts and other things" and arranged to be filmed delivering a speech at the Hudson County Fair. Continuing in a classic New York City vein when treating of the rural hinterlands, the *Times* reports that Heart's plan was to send "reliable agents" to "the out-of-the-way places, where a real campaign speech is rarely heard, even in a Presidential year, and where the farming population, practically cut off from all contact with the outer world after the last Summer boarder has left, will gladly drive many miles to listen to a talking machine and see a moving picture show." Hearst's agents were to offer their oratorical show in local halls, or "where there is no hall, Mr. Hearst's agents will set up the graphophone in a corner grocery and turn on a Hearst speech whenever the village lights have tired of eating raisins." They also conceived the idea of circulating the recordings on a kind of lending library basis, as many farmhouses already had talking machines "which are kept to furnish entertainment in the long Winter nights by rendering the latest vaudeville hits." The *Times* of October 28 reports a trial run showing in Irvington-on-Hudson, which, the Hearst campaign boasted, "evoked almost as much enthusiasm as Mr. Hearst himself would have done." Apparently, "the mechanician who ran the show had to let the talking machine repeat Mr. Hearst's speech and the biograph da capo its entire performance." He did the show again at the train station and several times in the smoking car on his way back to the city. The article goes on to note the tour schedule for five of "the canned speech outfits" as they fanned out over the state (3).

Hearst, the mass media innovator par excellence, is exploiting here the perceived capacity of the new communicative technologies, sound recording and moving pictures, to extend the immediacy of a platform event involving a gathered, co-present public to a dispersed public that is beyond the reach of the interaction order. To the living voice of the sound recording, the film representation adds the gestural movement of the living body. And more: interestingly, the film was four minutes longer than the speech recording (ten minutes to the sound recording's six), so the film, as reported by the *Times*, "will not only show Mr. Hearst in the act of delivering his speech, but will exhibit the hand-shaking scene that followed. Mr. Hearst will be seen entering his carriage. The pictures will pursue that carriage to the station and then show the Hearst special train pulling out with the multitude giving him an ovation" (NYT Oct. 10, 1906, 3). So, continuing in the terms provided by Goffman (1983), the Hearst media offers to the dispersed audiences aspects of the spectacle as well as the game, eliciting their participative engagement in the mediated performance.

Note again here the capacity for simulation that arises out of the conjunction of the two media, sound recording and film. The verbal text, recall, was recorded in a studio in New York City, while the delivery of the speech and its associated activities were filmed at the Hudson County Fair. When the two were combined, however, the speech was rekeyed: the spectators were induced to connect the speech they were hearing from the phonograph with the one they were seeing on the screen, perceiving them as complementary facets of a single event. The two media together conveyed an even stronger sense of immediacy than either alone could accomplish even if the lack of synchronization meant that any correspondence between the individual phrases and gestures was lost.

In response to Hearst's media initiative, participants in Hughes's campaign devised a counteroffensive that also exploited the capacities of sound recording. The leader of the Lower East Side Hughes organization, Mayer Schoenfeld, announced a plan to deploy twenty-five wagon-mounted phonographs throughout the eastside area, accompanied by brass bands to play in the intervals between speeches. "Each phonograph will get a permit to address its audiences in public meetings," the *New York Times* reported (NYT Oct. 22, 1906, 3), but the speeches to be "reeled off" by the phonographs were not those of Hughes himself, who said he knew nothing about the project and expressed no opinion on it, but campaign addresses presented by his supporters in Yiddish and Russian.

The first trial run of the Lower East Side media blitz was apparently no great success. A brief article in the *Times* records that

> Two brief speeches were reeled off in English to small and unappreciative audiences. The first was entitled "A Voice from the Ghetto."

"Gracious!" exclaimed an old man. "I hear speeches, but I don't see no mans." Then he hastened away. (NYT Oct. 24, 1906, 1)

Apocryphal though this anecdote may be, it reminds us of the strangeness, in those early years of sound recording, of hearing a disembodied voice. Schoenfeld still predicted success for the effort once recordings were offered in Yiddish, but there seems to be no further record of this initiative, and we suspect it was abandoned.

This abortive project is noteworthy nevertheless, alongside Hearst's effort, for what it reveals about the nascent uses of sound recording technology in the construction of political publics. Whereas Hearst was exploiting the potential of recorded sound to disseminate his voice over large spaces by deploying multiple reproductions of his speeches throughout upper New York State to reach "the farmer vote" (NYT Oct. 28, 1906, 2), the English-speaking constituency conceived as a core element in the American polity, Schoenfeld and his anti-Hearst allies in the Hughes campaign envisioned a similar strategy in going after the urbans ethnic "ghetto" vote: make multiple recordings and disperse them more locally throughout the Lower East Side. Key to this effort was the recognition that these sectors of the electorate were most effectively addressed in their own languages. Interestingly, in spinning the rather unspectacular results of his initial experiment with recorded speeches to the *Times* reporter, Schoenfeld suggested that in the next phase, "he expected to create a sensation and win votes for the Republican ticket when his talking machines turn loose in Yiddish on residents of the east side letters written by Jacob H. Schiff and Oscar Straus to the Jewish Daily News" (NYT Oct. 24, 1906, 1). That is to say, Schoenfeld apparently saw the phonograph as a means of recontextualizing and reanimating political discourses cast in another genre and composed for print—letters to the newspaper—in a living voice, and not the voice of their authors, at that. His vision of the new medium was still closely tied to the old medium of print journalism, but it recognized the ethnic and linguistic heterogeneity of the polity. Both efforts, Schoenfeld's and Hearst's, targeted constituencies in terms of their distribution in space, Schoenfeld's more local, Hearst's more broadly—if still regionally—dispersed.

THE PRESIDENTIAL CAMPAIGNS OF 1908 AND 1912

We see the same processes even more strongly at work in connection with the elections of 1908 and 1912, the next point at which commercial recordings figure in presidential campaigns. Between May and September 1908, all three major companies—Edison, Victor, and Columbia—issued recordings by William Jennings Bryan, the Democratic candidate, and William Howard Taft, his Republican opponent. In the campaign of 1912, Edison recorded only Theodore Roosevelt, candidate of the breakaway Progressive Party, whom Edison himself supported, while Victor issued recordings of all three candidates: Roosevelt, Taft, and the Democrat Woodrow Wilson. The 1908 and 1912 recordings represented a new departure: political speeches of great situational immediacy, keyed to an impending election, addressed to "burning topics," as one advertisement put it, recorded by the candidates themselves and available for home consumption in mediated, commoditized form. The critical point here is that the commercial recordings were themselves part of the campaign process; citizenship was assimilated to consumerism and to the marketing of the phonograph as a medium of home entertainment. "You should buy both Taft and Bryan records," says an advertisement in the Edison Phonograph Monthly, a house organ for Edison dealers,[8] "and compare them in your own home and to entertain your friends" (EPM Sept. 1908, 15). Another ad, for the Victor recordings, states, "The Victor makes no comment on the political situation, but merely offers the views of the candidates, so that each citizen may be helped to a wise and intelligent decision"[9] (Voices 2000, 22). Campaign speeches framed as consumer commodities and as entertainment: these are formative moments in the development to which Habermas ([1962] 1989, 195) alerts us, when "private enterprises evoke in their customers the idea that in their consumption decisions they act in their capacity as citizens."

Edison, ever attentive to economic payoff, was explicit about the element of commoditization and his desire to reach a mass market with his company's recordings. The Edison Phonograph Monthly kept up a constant barrage of sales promotion ideas from June to December 1908. The trade journal offered a steady stream of potential ad and display copy:

> Don't Talk Politics
> Get a Taft or Bryan Record and Let It Do It for You
> 35¢
>
> Taft or Bryan?
> Edison Records with the Speeches of Both (EPM Sept. 1908, 10)

Edison's optimistic projection to his dealers was that "the Bryan Records should go a long way towards offsetting the present trade dullness" (EPM July 1908, 6). Nor did the marketing hype end with the election. After the election was over, the campaign speech recordings were rekeyed from the time-bound

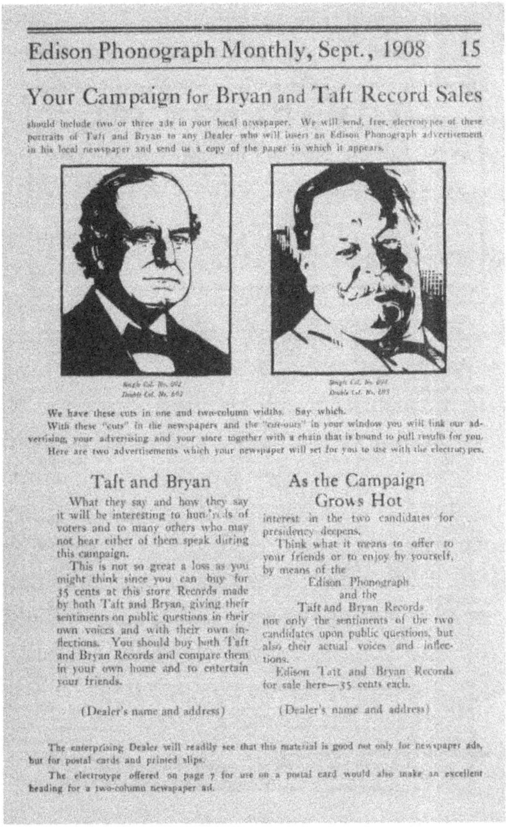

Edison Phonograph Monthly, Sept., 1908 15

Your Campaign for Bryan and Taft Record Sales

should include two or three ads in your local newspaper. We will send, free, electrotypes of these portraits of Taft and Bryan to any Dealer who will insert an Edison Phonograph advertisement in his local newspaper and send us a copy of the paper in which it appears.

We have these cuts in one and two-column widths. Say which.

With these "cuts" in the newspapers and the "cut-outs" in your window you will link our advertising, your advertising and your store together with a chain that is bound to pull results for you.

Here are two advertisements which your newspaper will set for you to use with the electrotypes.

Taft and Bryan

What they say and how they say it will be interesting to hundreds of voters and to many others who may not hear either of them speak during this campaign.

This is not so great a loss as you might think since you can buy for 35 cents at this store Records made by both Taft and Bryan, giving their sentiments on public questions in their own voices and with their own inflections. You should buy both Taft and Bryan Records and compare them in your own home and to entertain your friends.

(Dealer's name and address)

As the Campaign Grows Hot

interest in the two candidates for presidency deepens.

Think what it means to offer to your friends or to enjoy by yourself, by means of the

Edison Phonograph
and the
Taft and Bryan Records

not only the sentiments of the two candidates upon public questions, but also their actual voices and inflections.

Edison Taft and Bryan Records for sale here—35 cents each.

(Dealer's name and address)

The enterprising Dealer will readily see that this material is good not only for newspaper ads, but for postal cards and printed slips.

The electrotype offered on page 7 for use on a postal card would also make an excellent heading for a two-column newspaper ad.

Fig. 1.1. Advertising copy for Edison Phonograph dealers promoting the Taft and Bryan recordings, *Edison Phonograph Monthly*, September 1908.

topical urgency of "burning issues" to collectors' items, capitalizing on the aura of the presidency. We have more to say on this postelection marketing strategy in the conclusion (EPM Dec. 1908, 6).

In tacit acknowledgment of the restrictions on length imposed by the medium, Edison ads also make explicit that the recordings offer "selections" or "telling passages" (EPM June 1908, 6) from the candidates' speeches, but emphasize nonetheless their mimetic fidelity: "You can hear not only the exact words, but the exact tone and inflection of each Presidential candidate as he makes his speeches . . . each one a life-like representation" (EPM Sept. 1908, 15). Together with claims such as these, however, emphasizing the transparency of the medium—its immediacy, if you will—we find other statements that make a point of the technological mediation of the recording process, noting, for example, that "these records, the first ever made by THEODORE ROOSEVELT, were prepared with great care by our recording experts who have

successfully brought out the forceful and convincing logic of his arguments."
In an allied vein, a 1908 Victor ad for the recordings of Taft's speeches states,
"William H Taft Speaks to the American Public through the Victor" (Voices
2000, 8), neatly summing up the essence of the innovation, focusing on speaking, the communicative medium of co-presence, but here addressing the dispersed American public through the mediation of the Victor talking machine
recording.

One interesting feature of the reproduction and recontextualization of spoken oratory was that the process rendered the deictic center of the utterance
ambiguous. Traditional oratory is always conspicuously sited in public spaces
and scheduled in the program of public, collective events: these are defining
attributes of the kinds of cultural performances in which political oratory characteristically occurs. The deictic calibration of recorded oratory was rendered
problematic in a number of ways; we deal with a further aspect of this dynamic
later on. It will be of interest here, though, to suggest how deictic ambiguity
entered into the framing of the campaign records, specifically in the devising
of advertising copy. Among the ideas for window posters suggested in the *Edison Phonograph Monthly* are "Bryan Speaks Today" and "Taft Speaks Tonight
in an Edison Phonograph" (EPM Sept. 1908, 10). A homemade window sign,
sent in by one of the dealers, adds some paronomastic ambiguity to the deictic
ambiguity:

> Come in <u>Here</u>
> And <u>Hear</u>
> Them Speak!
> Who?
> The 2 Bills

That is, William Jennings Bryan and William Howard Taft (EPM Nov. 1908,
20). When is the "today" or "tonight," and where is the "here" for the campaign
speech in the candidate's "actual voice"?

This deictic play with immediacy and mediacy was all part of the larger
process of making sense of the new medium as a vehicle for campaign oratory.
The various framings and applications are illuminating. Consider this account
from the *Washington Post*:

> Two hundred and fifty Washington Democrats listened last night in the local
> club's quarters ... to a number of speeches by Mr. Bryan, delivered through
> the medium of the phonograph. . . . Much enthusiasm was displayed, but no
> applause interrupted the speech of Mr. Bryan. All of it came after the speech
> was finished.

> The topics on which Mr. Bryan addressed the audience were "The railroad question," "The trust question," "The Guarantee of bank deposits," and "Immortality."[10]

At the beginning of the account, the mediation of the phonograph is explicit, but in the second paragraph the mediating technology drops out, and it appears that "Mr. Bryan addressed the audience" directly. This report is also interesting in what it suggests about audience responses to the recorded speeches. The default expectation for live speechmaking was that audience members had license to interrupt the orator with applause, whereas this audience, at least, saved its applause—however enthusiastic—for the end. This shift in response patterns may have been a consequence of the lack of obvious applause lines in the recorded speeches. Note also that in the event that listeners might have been moved to applaud, the speaker would not have paused to accommodate them but would have continued to speak right through the applause.

In addition to political club meetings, there were other public frameworks into which recorded speeches might be employed to stand in for the living candidates. Consider, for example, what was billed as "The First Phonograph Debate in History," held in Des Moines, Iowa, on October 9 and conducted by means of Edison phonographs playing speeches by Bryan and Taft in alternation.

> The Opera House was packed with an audience of 1,500 persons, all of whom seemed much pleased with the affair. The machines were plainly heard in all parts of the house. The debate was interspersed with vocal and instrumental music by local artists. At the close of the affair a number of miscellaneous selections were played on the Phonograph, including some of the Amberol Records. The event was voted a great advertisement for the Edison Phonograph. (EPM Nov. 1908, 5)

This event has all the features and accompaniments of a full-scale political debate: held in a large public auditorium before a sizable gathering of people, featuring live music as well as recorded selections, and so on. This "debate," like the earlier political club meeting, is just the kind of event in which audiences were accustomed to hearing campaign speeches; the only difference, again, is that the phonographs stood in for the living candidates—and, of course, that it was not only a political event but an advertising event for the Edison phonograph.

These conventional platform contexts for political oratory provided the basis for imagining a recontextualization of oratory from the public event to a new space—the "here" of the ad poster quoted earlier—and a flexible time: "today," "tonight," "at your convenience" (EPM July 1908, 14), any time the recordings are played. A pair of 1912 Victor ads captures especially effectively the ambiguous and emergent understandings of this new communicative technology vis-à-vis political oratory, poised between a visionary imagining of its unique capacities on the one hand and a conservative framing of its representations on the other.

> Would you accept a special invitation to hear Mr. Taft, Mr. Wilson, and Mr. Roosevelt speak from the same platform? Then come in and hear them discuss the important topics of the campaign, just as you would hear them if seated in a convention hall with these three great men speaking to you. (Voices 2000, 21)

> The Republican, Democratic and Progressive candidates have decided to present their views to the people through that greatest of all public mediums, the Victor, which will bring directly into the home the actual voices of the aspirants for Presidential honors.
>
> Heretofore, only a very small proportion of the people were able to listen to the candidates in person. Now, for the first time in the history of our country, the Victor makes it possible for the people to hear the actual voices of the three nominees in a discussion of the principles involved in the campaign. This debate, an intensely interesting one, fills eighteen records, most of which have been combined in double-faced form, thus insuring the widest publicity for the discussion. (Voices 2000, 22)

Both advertisements are exercises in virtual reality, setting up for an undifferentiated mass of potential customers conditions in which those who accept the invitation to buy the records will be virtually transported from the dispersed settings of the record dealer's shop or their individual homes—listening to the technologically mediated, disembodied, and fragmented voices of the separate candidates—into the selected, gathered audience at a live political debate. As a member of that select audience, you are the directly targeted addressee of the great speaker's words. The force that actualizes this complex virtual reality is "the actual voices" of the candidates, mediated though they are through the Victor talking machine, "which will bring into the home the actual voices of the aspirants for Presidential honors" (Voices 2000, 22). You might even imagine that the presidential candidate himself has come to your home.

The model advertising letter offered in the *Edison Phonograph Monthly* in July 1908 sets up just such a virtual scenario: "If William Jennings Bryan offered to deliver his favorite orations in your home, you would consider you had a very great privilege, would you not? Well, we make you an offer that practically amounts to the same thing" (EPM July 1908, 9). The power of presence embodied in the voice is the pivot-point around which the new experience of hearing campaign speeches in the shop or in the privacy of domestic space is assimilated back to the more familiar—if less widely accessible—experience of listening to campaign speeches by live political luminaries in convention halls. Now the reader of the ad in the first example is asked to imagine himself or herself back in public space. But interestingly, the ad turns at the end from this imagined restoration of the speeches to the context of a live debate to invoke a dispersed, distributive public, for it is through the diffusion of these recordings that "the widest publicity for the discussion" can be achieved. The reader of the ad in the third example is asked to imagine that the public orator has come to his or her home. Thus, although the distribution of campaign speeches by commercial recording shifts the venue from large-scale public events to the intimate domestic setting of the home, it can be viewed as widening public access to the political process. Formerly, the ad argues, "only a very small proportion of the people" could hear the voices of the candidates, notwithstanding the scale of the gatherings in which they delivered their speeches, whereas now "the people"—the implication is, all the people, the polity—can hear them on recordings, provided, of course, that they can afford the records and the machines to play them. "Bryan Speaks to Millions through the Edison Phonograph" proclaims an *Edison Phonograph Monthly* ad (EPM July 1908, 18). This is very much an expansion of the bourgeois public sphere.[11]

Unfortunately, it is hard to assess how and to what extent purchasers used the campaign recordings in the privacy of their homes. We can find only sporadic references to them being played during private social entertainments, as for instance: "Mr. and Mrs. John Henry Lynch entertained a small company of friends at their home on North Buchanan street [in Edwardsville, Illinois] last night. Refreshments were served and a social time enjoyed. The particular feature of the evening was the rendering of the ... campaign records on the graphaphone [*sic*]."[12]

There is also a gendered component to the construction of a dispersed public by means of commercial recordings. In the lead-up to the 1908 election, the Edison company several times makes a point to its dealers that the Bryan records "will appeal very strongly to women as well as men" (EPM June 1908, 6): "Don't make the mistake to think that men are your only possibilities. Far from it. Women flock to hear Bryan whenever he speaks. It takes a large 'men only' sign to keep them away, and even then they do not stay away through choice. . . . Indeed, you cannot afford to overlook the ladies" (EPM July 1908, 9). Here, there is at least the suggestion that women may have unrestricted access to political oratory through campaign recordings played in domestic space. Still years away from having the vote, and barred by Men Only signs from certain public events at which candidates hold forth, women may nevertheless be an important part of the commercial market for the recorded speeches as commodities, and that is what counted for recording industry entrepreneurs like Edison. It is still the case, however, that the recordings might bring their "men only" history with them, creating a disjunction between past and present contexts, as in the use of the vocative "Gentlemen" to open a speech.[13]

The recorded texts themselves likewise signal the ambiguity of a new medium whose capacities have not yet assumed—or been disciplined into—a clear shape. For example, the cylinder and disk formats available at the time allowed for recordings of around two and a half to four minutes in duration. This impelled the recorded speeches toward topical and formal closure within the relatively brief and bounded span of a single recording. One trade journal referred to the recorded speeches as "tabloid addresses," which, in contemporary usage, highlighted their brevity, concision, and concentration of expression (EPM June 1908, 6). An especially revealing newspaper account observes: "The records are short and directly to the point. They deal with the conspicuous issues discussed by the candidates in a simple and straightforward manner. Nobody wearies of listening to them. The different appeals are completed within the brief space of three minutes, and they make a strong and lasting impression on the mind. On any of the single questions treated by the phonograph records one gains as clear an idea as if he listened to a speech an hour or two in length covering all of the issues of the campaign."[14] Brevity is here tied to rhetorical efficiency: short speeches get the point across without tiring or boring the listener. Pulling in the other direction, however, were the generic expectations of the campaign speech, which tended to be considerably longer and more complex, both in argument and form.

The candidates seem to have negotiated this tension in different ways. Notwithstanding the central linkage of the recordings to the presidential campaign, two of Bryan's ten recorded speeches were oratorical display pieces lifted from earlier speeches to enhance the attraction of the project to potential customers. The remaining eight recordings featured speeches that Bryan composed expressly for the recordings, drawing themes and occasional phrases from his longer speeches. These new speeches addressed "the leading political questions of the day" and were closely keyed to the principal issues with which Bryan had long been publicly identified and to major planks in the Democratic platform. Of "Imperialism," for example, the *Edison Phonograph Monthly* notes, "This is a subject on which Mr. Bryan never fails to delight his hearers" (EPM June 1908, 7). Bryan's issue-oriented speeches are rhetorically tight. They begin characteristically by defining the principal elements of the problem at issue or identifying an existing condition that serves as a frame of reference for the delineation of his own position. Following this framing section is generally a statement of what is required to answer or remedy the problem and Bryan's or the party's policy to provide the necessary corrective and the moral grounds for doing so. The speeches are predominantly expository and deliberative, even didactic; one journalistic observer suggested that "the new use to which the phonograph has been put gives it a respectable rank as an educator."[15] Poetic devices and virtuosic flourishes are conspicuously minimized. If the recorded speeches are far from full-blown virtuosic displays, they are also, by their mediated and semiotically reduced nature, restricted in their capacity to invoke the strong affect that in-person performances might achieve. For one thing, while the recorded speeches evince intertextual links to Bryan's longer campaign speeches, both from the current and earlier campaigns, they are conspicuously free of other contextualization cues (Gumperz 1982, 131): no vocatives of address to hearers, no deictic anchorings in time or place, no indexing of situational contexts of production or reception. The absence of contextualization cues, to be sure, rendered Bryan's recorded speeches maximally open to insertion into new contexts, like the public platform events we described a few pages back.

Taft, on the other hand, expected to draw extracts directly from his longer speeches, though according to the *New York Times* (NYT Aug. 3, 1908, 3), he did rehearse beforehand. The article goes on to note that Taft's preparation also included listening to one of the recordings made earlier by Bryan. After an uncertain start, Taft apparently really got into the process and wound up recording a total of twelve speeches instead of the four originally intended. One of the extras was an after-dinner speech on "Irish Humor" that is largely a corny travel account of his trip to Ireland with a declamation of the popular recitation piece "Shandon Bells." While this may well have been a bid for the Irish vote, it represents an instance where the appeal of oratory as entertainment supersedes the topical and rhetorical urgency of the campaign for office.

Given Taft's modus operandi of snipping excerpts from his longer speeches, it is not surprising that there are instances on the campaign recordings where the disassembly of longer speeches into short, bounded, and finalized units is imperfectly accomplished, leaving traces of the cohesion that tied the original text together. For example, the Taft recording entitled "Republican Responsibility and Performance; Democratic Responsibility and Failure" (Columbia 14503; https://purl.dlib.indiana.edu/iudl/media/d375844c35) begins with this parallel sequence: "I have already pointed out that the Republican party long ago passed the Antitrust Law and is vigorously enforcing it. I have already stated that it passed the Interstate Commerce Law and its amendments, the

Elkins Law and the Rate Bill, and is vigorously enforcing them. I have already dwelt on the great change for the better that has been brought about by this administration." As this is the beginning of the recording, Taft has not "already" done anything. Nor does this deictic adverb point to any of his other 1908 recordings. The recorded speech comes from a much longer speech—the recording amounts to about 13 percent of the whole—delivered at Hot Springs, Virginia, August 21, 1908, and occurs about midway through the text, after Taft has indeed "already" said the things he indexes in this clip (Taft 2001, 1–30).

Also revealing is the deictic alignment of the recontextualized speeches to situational contexts of utterance as well as to co-text. Consider, for example, the following passage from another 1908 Taft recording, "Foreign Missions" (Edison 9996; https://purl.dlib.indiana.edu/iudl/media/x31q47zg9p): "I am not here tonight to speak of foreign missions from a purely religious standpoint. That has been done and will be done. I am here to speak of it from the standpoint of political governmental advancement" (Debate 2008, 36–37, track 3). What time and place are indexed by "here tonight"? The recorded utterance has carried some of its history with it in the process of recontextualization from the gathering at which it was originally spoken—the referent of "here tonight"—to the recording session, and beyond that to each playing of the record. This marks it as a reiteration of words originally spoken at another time and place, even if the author/speaker is the same individual.

To take another example, a 1912 recording by Theodore Roosevelt, entitled "Why the Trusts and Bosses Oppose the Progressive Party" (Victor 35250a, https://purl.dlib.indiana.edu/iudl/media/465544j32k), opens with the sentence "Now this statement of Mr. Archbold represents but part of the truth." "Now this" appears to be a double deictic, but where is it anchored? "Now" actually serves here as a discourse marker signaling a transition in an ideational sequence and is therefore anomalous at the beginning of an utterance such as this with no antecedent co-text; the demonstrative adjective "this" demands an antecedent as well. As it happens, though, the preceding recording by Theodore Roosevelt, "The 'Abyssinian Treatment' of Standard Oil" (Victor 35249b; https://purl.dlib.indiana.edu/iudl/media/b98m414b3t), does introduce a statement by Mr. Archbold of Standard Oil, and the "Now this" of the recording at hand expresses a cohesive link that was fully motivated in the original, unified text.

The point is that the campaign recordings were unsteadily poised between varying alignments to an audience and other aspects of context; they are unsure of their footing in Goffman's sense. Much of the work of contextualization is devoted to negotiating the transition between the gathered, co-present, co-participant public of those events in which political speeches were conventionally delivered, addressed directly to the assembled audience, and the dispersed public of record buyers, sited in private, domestic space, listening to commoditized speeches for which the targeted addressee was not clear, by an absent orator, who was nevertheless still somehow present, through his voice.

CONCLUSION

The experiment of disseminating campaign oratory via sound recordings was short-lived. No campaign recordings were made for the election of 1916, and the campaign of 1920 marked the advent of radio, which opened an entirely new chapter in the relationship between politics and media. Brief though it was, however, the recording of campaign speeches represents an arena in which some of the basic work was done in effecting the transition from political oratory as a means of constituting an assembled public to the mediated political address of radio and TV in relation to the dispersed political public that is a feature of contemporary political life. Habermas ([1962] 1989, 162) suggests that "the bourgeois ideal type assumed that out of the audience-oriented subjectivity's well-founded interior domain a public sphere would evolve in the world of letters. Today, instead of this, the latter has turned into a conduit for social forces channeled into the conjugal family's inner space by way of a public sphere that the mass media have transmogrified into a sphere of culture consumption." The examination of how political oratory was adapted to early commercial sound recording suggests how this transformation was initiated. In particular, we have examined the shifts in alignment to a public that attended the recontextualization of political oratory from live performance to sound recording.

A key outcome of this process, we have argued, was the refiguration of an oratorically constituted public from what we might identify as a gathered, assembled, co-present public, characteristic of the large-scale cultural performances—political rallies and other ceremonial occasions—that were the conventional context for political oratory to the dispersed public, consisting of individuals and strong groups listening to political speeches on commercial recordings in private, domestic space. Although this recontextualization was imagined very early, almost at the moment when the phonograph was invented, the process was gradual and experimental, as the developers of the new technology of sound recording tried to figure out the capacities of the new medium and imagine how to exploit them.

Clearly, the contextual association of oratory with large-scale, gathered events remained salient, such that producers of recorded speeches felt the need to preserve elements of the default context as an orienting framework to evoke the immediacy of the performance event. Paradoxically, this impulse toward immediacy and verisimilitude required dissimulation and illusion, as in the inclusion of applause—provided by recording studio personnel—on the recordings of speeches from the 1896 presidential campaign, and the ambiguation of the animators of those speeches by omitting the opening announcements of the performer that conventionally framed early recordings.

A complementary way of retaining the contextual expectations of political speeches on record was to use the recordings as surrogates for live speakers within the same kinds of gathered events that represented the default context for political oratory: campaign rallies, political meetings, and the like. The speeches were disembodied in the recording process but then reinserted in conventional contexts for oratorical performance.

Yet another demonstration that the established orienting frameworks for the production and reception of political oratory had staying power was Hearst's coupling of his recorded speeches with films of a larger event featuring political oratory, the Hudson County Fair, including visual recording of the other constituent events surrounding the oratorical performance, such as shaking hands with members of the crowd, the departure of the candidate for the train station, and so on. And even as the record companies targeted the commercial recordings of campaign speeches by the presidential candidates in the elections of 1908 and 1912, they invited potential consumers to imagine themselves as participants in gathered public events, enjoying the immediacy of co-presence with the speaker that the preservation of his "living voice" on records made possible.

At the same time that the producers of campaign recordings worked to invest them with contextualization cues that continued to align the recorded speeches to a gathered public, the speeches themselves carried elements of their originary context with them: deictic references to time and place, vocatives, cohesion devices—a special problem because of the temporal constraints of the recordings—and the like, which rendered their footing uncertain and ambiguous.

The element of commoditization itself contributed to the alignment of the recorded speeches to a public. The casting of the public as also a market gave the producers of the recordings an interest in maximizing the circulation of the recorded speeches through sales of the records and the machines on which to play them. Thus, for example, while women were not included in the electorate, they could certainly be included among the consumers of the records as commodities. In general, the potentially broad circulation of campaign speeches on records beyond the limits allowed by the carrying capacity of live political events, while it might be cast as a contribution to civic inclusiveness, was in fact conceived in terms of the cultivation of the bourgeois market, the constituency with the economic means to buy the recordings and phonographs. Also market driven was the reframing of the campaign recordings from timely discourses on current and pressing political issues, keyed to the ongoing campaigns of 1908 or 1912, to the durability—even timelessness—of collectibles, after the elections were over.

An entry in the December 1908 issue of the *Edison Phonograph Monthly* (EPM Dec. 1908, 6) suggests that "with the passing of Election Day the Edison Records made by Mr. Taft and Mr. Bryan take on an added value, and enterprising Dealers will not be slow to take advantage of it." Taft's records will be "the first in the history of the world to be made by the head of a great government." Taft, as president, will become an even larger figure than he was as a candidate, operating not only on a more prominent national stage but also before "the entire world." Once he assumes office, then, "the owner of one of these Records can say with no little pride: 'This is a Record made by the President of the United States.'" Campaign issues will recede in importance once Taft is president. Rather, the entry suggests, "the Records of his *voice* will become more and more interesting, and they will have a value that could not be estimated if it were not possible to duplicate them." This foregrounding of the president's *voice* hearkens back to the earlier framing of the campaign records as a means of bringing the candidates into your own home, their voices serving as instruments of presence. At the same time, this marketing aid reminds the reader that commercial records are multipliers of value, allowing all buyers of the Taft records to share in the great value of owning a piece of the president's voice insofar as the duplicating power of recording technology makes his speeches accessible to the broad public of record buyers.

But what of the losing candidate? "The Bryan Records," the entry contends, "will lose none of their selling power because Mr. Bryan has again been defeated for the Presidency. Mr. Bryan is regarded by his friends and opponents alike as one of the most popular men and greatest orators this country has produced, and it will always be a pleasure to listen to the Edison Records made by him." Here, though, the marketers concede that Bryan's issue-oriented speeches may well be of declining interest to prospective buyers as their timeliness wanes. But the two recordings drawn from earlier speeches that were proven crowd-pleasers, "Immortality" and "An Ideal Republic," added to flesh out the collection, "are not political in any sense, but are masterpieces of oratory and diction" and so "must be sought after for years to come." Buyers will want to own them for the pleasure of hearing Bryan perform.

Ultimately, then, what we are offering in this chapter is a historical window on a communicative technology still in its formative stages, somewhat inchoate, open to imagination, and still keyed in significant part to antecedent expectations, orienting frameworks, and practices for the production, repetition, and circulation of discourse at the same time that it anticipates a highly consequential refiguration of the public sphere. And there is, we should say, far more to the story: the campaign speeches we have discussed were but one representation of political oratory on early commercial sound recordings, and our engagement with them is part of a more comprehensive interest in sound recording and the formation of political publics that will be a recurrent theme in the chapters to follow.

NOTES

1. Hereafter cited as NYT.

2. Wendell Phillips was one of the most celebrated orators of his day, an abolitionist and champion of the rights of women, organized labor, and Native Americans. See Bartlett 1962 and J. Stewart 1986.

3. For an allied approach, see Lempert and Silverstein 2012, 32.

4. "Congressman Filkin's Home-Coming," descriptive talking by Byron G. Harlan, orchestra accompaniment, United A1036, mx. 19235 (https://purl.dlib.indiana.edu/iudl /media/n79h14hf1n).

5. George Graham, "Free Silver Orator" (Berliner 660, November 30, 1896, https://purl .dlib.indiana.edu/iudl/media/9504952734):

Announcement: Talk on silver, by George Graham.

MC: Friends, I appear before you this evening to introduce the honorable Erastus
    Silverspotter from Browntown County, Kansas.

Silverspotter: Friends, I am here

and I have come this evenin'

to speak to you empty hats of the working man.

The working man has stood it long enough!

And he will stand it no longer!

And I say that you shall not press down upon the brow of labor this crown of thorns!

Neither shall you crucify mankind on a cross of gold!

And friends, my first legislative act will be, if I am elected, is to have the price of liver
    reduced to six cents per pound!

Then we will load up express wagons with free silver and distribute it to everybody free
    of charge!

That's what we want.

6. "William McKinley—1896 Campaign Speech," YouTube video posted by reaganwayne, June 7, 2007, 00:01:04, https://www.youtube.com/watch?v=m6ZUneyU7Vo.

7. Accessed October 15, 2020, http://phonozoic.net/primtexts/n0065.htm.

8. Hereafter cited as EPM.

9. *In Their Own Voices: The U.S. Presidential Elections of 1908 and 1912.* Marston CD 52028-2, 2000; hereafter cited as Voices.

10. "Listen to Bryan Records, District Democrats Supplement Real Speeches by 'Canned' Ones," *Washington Post*, September 4, 1896, 3.

11. For a discussion of the scalar implications of distributing political speeches via commercial recordings see Bauman 2016.

12. *Edwardsville (IL) Intelligencer*, September 5, 1908, 1.

13. William Howard Taft, "Roosevelt Policies," Edison 10002. *Debate '08: Taft and Bryan Campaign on the Edison Phonograph* (2008, 32–33, track 1). Hereafter cited as Debate.

14. "As to 'Canned' Oratory," from *Kansas City Star*, in *Washington Post*, September 8, 1908, 6.

15. "As to 'Canned' Oratory," 6.

TWO

"ACCORDIN' TO THE GOSPEL OF ETYMOLOGY"

Aural Blackface and New African American Poetics

INTRODUCTION

Beside the "bottled" orations in the speech connoisseur's cellar, imagined in the *New York Times*'s early journalistic effort to grasp the potential uses of Edison's wondrous invention, the writer foresaw a rack of vintage sermons as well. Sermons, like orations, were elevated speech forms, vehicles for the display of virtuosity as well as virtue and thus clearly worthy of storing up for future consumption. "A large business will, of course, be done in bottled sermons, and many weak congregations which are unable to pay a regular Pastor will content themselves with publicly opening a bottle of 'Dr. Tyng,' 'Dr. Crosby,' or some other popular ministerial brand" (NYT Nov. 7, 1877, 4). The playful prophecy of the newspaperman who foresaw the branding of bottled speech performances proved to be remarkably prescient. Two decades after he published his speculations, as the commercial record companies vied with each other in building their catalogs, their aim was indeed to establish their brands as the frame of reference for the consumer market they were struggling to create. When it came to the place of sermons within those developing catalogs, however, celebrity preachers occupied a limited place. Here too, though, the early commentator proved prescient. The playful and irreverent tone of his speculations about the declining need for living preachers in organized religion anticipated the conspicuous irreverence of the recorded sermons offered to consumers in which parody was far more prevalent than piety.

The most common sermon parodies in the early commercial record catalogs had a special edge. They were burlesque performances, drawn from the minstrel-show tradition, in which performers in blackface animated popular

40

stereotypes of the traditional African American preacher. Having devoted the preceding chapter to political oratory in a serious key with a couple of side-glances toward parody, I devote this chapter to these parodies of the African American oral sermon and what we can learn from them about the popular entertainments of the day, the emergent culture of commercial sound record-ing, and the racial politics of the US at the turn of the twentieth century.

The African American sermon and the preaching that gave it expression were manifestly complex ideological signs in the politics of race in early twentieth-century American culture. These expressive forms and practices were bound up in the construction of and contestation over the place of Black people in Ameri-can life, past, present, and future. For African Americans, what was at issue were the debilitating and degrading stereotypes that encoded purported traits that marked them as inferior, ignorant, and incapable of cultural or economic or intellectual achievement. The stereotypes were largely White constructions, but they exerted enormous hegemonic force over White and Black people alike.

One of the most resonant and provocative contributions to the effort on the part of the African American intellectuals of the Harlem Renaissance of the 1920s and 1930s to grapple with the place of vernacular preaching in the history of their people and to imagine where it might fit within their emergent visions and programs for the future is a slim volume of poetry called *God's Trombones: Seven Negro Sermons in Verse*, published in 1927 by James Weldon Johnson (1976). "God's trombone" was Johnson's evocative term for the voice of the African American preacher, "the instrument," he wrote, "possessing above all others the power to express the wide and varied range of emotions encom-passed by the human voice" (J. Johnson 1976, 5). Johnson's motivation for writing *God's Trombones* was rooted in the Romantic ideology of many intel-lectuals of the Harlem Renaissance that the literary and artistic creations of a people represented a powerful—for Johnson, the most powerful—basis for establishing the greatness of their culture and sustaining a claim to respect from others. Johnson (1931, 9) argued, for example, in 1931, that "no people that has produced great literature and art has ever been looked upon by the world as distinctly inferior."

For African Americans, as we know all too well, the burden of being looked down upon was especially severe, so a claim to being the creators of great lit-erature and art required a strenuous effort to overcome prejudice and negative stereotypes. In the trenchant words of Alain Locke ([1925] 1992, 264), one of the leading intellectuals of the Harlem Renaissance, it was incumbent upon participants in the movement to "discover and reveal the beauty which preju-dice and caricature have overlaid."

Johnson was clearly committed to that task of discovery and revelation. In his view, "the old-time Negro preacher has not yet been given the niche in which he properly belongs," as a powerful contributor to the verbal arts of African American people. "He has been portrayed only as a semi-comic figure" (J. Johnson 1976, 2). Johnson recognized in the "old-time Negro preacher" an orator who understood that oratory "is a progression of rhythmic words more than it is anything else. . . . He had the power to sweep his hearers before him; and so himself was often swept away. At such times his language was not prose but poetry." The oral poetry of the preacher, shaped by "a highly developed sense of sound and rhythm" and marked by such devices as "rhythmic intoning," "a high pitch of fervency," "a fusion of Negro idioms with Bible English," and "a certain sort of pause that is marked by a quick intaking and audible expulsion of the breath," was an inspiration to Johnson in his quest for a true African American poetic style: "It was from memories of such preachers there grew the idea of this book of poems" (3–8). It is worth noting that Johnson consistently refers to the preachers who inspired *God's Trombones* as "old-time." At the end of his preface, he observes explicitly that "the old-time Negro preacher is rapidly passing" (8). That is to say, in Johnson's view, the kind of preaching he finds so rich in poetry was, by the late 1920s, something of an anachronism but a useful resource for modern artistic adaptation.

In writing his poems, Johnson (1976, 5) had to struggle not only with the stereotype of the preacher as a comic figure but with issues of dialect, because of "the fixing effects of its long association with the Negro only as a happy-go-lucky or a forlorn figure." That is, dialect was a vehicle of trivialization or tragedy. In other words, Johnson was deeply concerned, in writing *God's Trombones*, with providing a critical corrective for strongly established negative stereotypes of the Black preacher and caricatures of his language. In the long run, Johnson was successful: his book has been a precious and vital resource for subsequent writers and scholars who have come to recognize the poetic power of Black preaching, from Zora Neale Hurston's (1934) *Jonah's Gourd Vine* (see also Hurston 2022, 51, 78, 80) on down.

Much less clearly known, however, regarding the language of Black preaching, is the base of stereotype and caricature against which Johnson addressed his book. There is some interesting work on burlesque sermons in antebellum minstrelsy (e.g., Holmberg and Schneider 1986; Mahar 1999, 59–86) but none on the late nineteenth- and early twentieth-century published texts—from the period more immediately antecedent to Johnson's work—which are found mostly in resource books for White amateurs who wished to put on a minstrel show (e.g., Dumont 1899; Marble 1893; Simond 1974). Those texts are indeed replete with crude and stereotypical representations of dialect, but they are of next to no use as sources of insight into how African American performance style was actualized on the minstrel stage. Given Johnson's insistence on the figure he persistently called the "old-time Negro preacher" as an oral poet, whose great artistic achievement was realized in the living context of the religious service, it would be useful to have a corpus of materials that sheds light on how the Black preacher was caricatured in performance. Unfortunately, we don't have recordings of full-blown minstrel shows. What gets us closest to enacted representations of the comic preacher is a body of commercial sound recordings from the first two decades of the twentieth century featuring parodic sermons drawn from the blackface tradition. The recordings, of course, rely only on sound, without the visual component of actual blackface; they are a sort of "aural blackface" (Strausbaugh 2006, 225). I suggest in the body of the chapter how Blackness is conveyed in these recorded performances. There are some interesting and surprising things to be learned from those recordings, and in the remainder of this chapter, I propose to offer a preliminary exploration of what they reveal about stereotypes of Black preaching.

Before taking up the mock African American sermons that are my main subject, though, let me establish two points of background. First, the burlesque sermon as a genre has a very long history in Euro-American tradition. The *sermon joyeux*, in which a lay performer delivered a mock sermon in full homiletic style on a decidedly earthy topic, was a well-established genre of the medieval carnivalesque (Gilman 1974; Jones 1997). A 1712 statute of the Massachusetts Bay Colony, outlawing the "composing, writing, printing or publishing of any filthy, obscene, or profane song, pamphlet, libel or mock sermon, in imitation or in mimicking of preaching, or any other part of divine worship," testifies to the continuation of the tradition on this side of the Atlantic and points up the generally antiauthoritative power of parody and burlesque irrespective of race (Anon. 1814, 399).

Moreover, the figure of the preacher, the embodiment and agent of religious authority, has always been an apt target for deflation when he slips from his pedestal. From the vantage point of verbal performance, the preacher is expected to be fluent in delivery and coherent in message, so any frame-breaking display of verbal incompetence is a ready resource for humor. There is an early recording by Cal Stewart, "A Revival Meeting in Pumpkin Center," reporting on the fictional, rural New England town that served as the setting for an extensive series of widely popular recordings that Stewart, one of the classic avatars of American "rube" humor, made in his performance persona of Uncle Josh Weathersby. We will see—and hear—much more about Cal Stewart in later chapters. One episode of Stewart's account of the revival meeting recounts the effort of Rev. Obadiah White to preach a sermon, taking as his text the well-known opening line from a poem by the great Irish poet, Thomas Moore ([1816] 1896, 147–148), "This world is all a fleeting show."

Excerpt from: **Cal Stewart, "A Revival Meeting in Pumpkin Center" (US Everlasting Indestructible Cylinder 1349-1; https://purl.dlib.indiana .edu/iudl/media/g05f36bs93), released 1909–1913**

Well, the Reverend Obadiah White was a'preachin' to us,
and he went to say,
"This world is but a fleeting show."
And he said, "'This world is but a flowing sheet.'
I should've said, 'This world is but a shoating flea.'
I mean, dear brothers and sisters,
'This world is but a fleeing shoat.'"
[Laughs]
Well, the choir had to sing four times
before they could get order, an'
I just had to snicker right out.
[Laughs]

What happens, however, is that the hapless preacher commits a series of enunciative misfires that subvert the moral import of his text. Attempting to say, "This world is but a fleeting show," he first produces "flowing sheet," then "shoating flea," and finally "fleeing shoat," with each subsequent metathesis representing a failed effort at repair (as in "I should have said," "I mean") of the preceding one. The result of this cumulative series of spoonerisms is uncontrolled, carnivalesque laughter, a breakdown of the reverent tone of the sermon, and a state of general disorder. In Stewart's recorded performance, then, the reported sermon serves entirely as a vehicle for a bit of entertaining speech play, stemming from a momentary breakdown. The performance is not a sustained assault on sermons or preachers in general.

Also by way of background, I want to offer to the reader an expressive baseline against which to hear the burlesque sermons. The first recordings we have of African American sermons date from the mid-1920s, as the commercial producers of so-called race records—recordings by Black performers, oriented to the emerging market of African American consumers—realized that there would be a sizable audience for virtuoso religious performers (Martin 2014). Recordings by these figures, of course, postdate the burlesque sermons that are the focus of this chapter. Still, they are close enough in time and expressively continuous with the traditional preaching of the period spanned by this book that they serve well to document the sermon style against which the blackface performers I discuss below framed their own performances. African American sermon recordings dating from the mid-1920s onward are easily found on the internet. I urge readers to search out recordings by such popular preachers as the Rev. J. M. Gates, Rev. A. W. Nix, Rev. F. W. McGee, and Rev. Leora Ross, the sole female preacher to be featured on commercial records. With the sound of the real thing in your ear, you will be better equipped to engage with my argument in the pages that follow.

GEORGE GRAHAM, "COLORED FUNERAL"

The first example of a burlesque sermon I consider is entitled "Colored Funeral," recorded in 1901—fairly early in the development of commercial sound recording—by George Graham, whom we encountered in his guise as a pitchman in the introduction to this book. Graham was certainly adept at the pitchman's art, as attested by his recordings of spiels for baking powder, liniment, a corn cure, and a carnival side show—all displays of verbal virtuosity—but he apparently commanded a much broader range of performance skills. He was known around the capital area as a blackface comedian, and his recordings also included various forms of oratory (both serious and parodic), comic narrative, Irish and German dialect humor, humorous sketches of various kinds, and other forms drawn from the contemporary repertoire of popular entertainments (Feaster 2007, 493–501).

George Graham, "Colored Funeral" (Victor M-982-1; https://purl.dlib .indiana.edu/iudl/media/b98m414b2h), March 7, 1903

Announcement:
Imitation of an old-time colored preacher down South,
buryin' one of de brothers,
by George Graham.

Sermon:
Now, my dear beloved brothers and sisters,
I want to say one thing dis mawnin',   5
dat in de midst of life,
we are in death.

Yes, and dat fact is forcibly brought to our minds
every day.
Every day you can see it on every hand.   10

You can see it in de mountains
and in de valleys.

A:::h, my dear beloved brothers and sisters,
prepare for dat mighty time t'come.
I want you a:::ll to prepare   15
for dat mighty time t'come.

Now, I am gathered here dis mawnin'
to perform a sad and painful duty.
One of our dear beloved brothers,
by de name of Flatback Jackson,   20
am no mo'.

He am done passed over dat dark river
from which no traveler ever is known to return.
And if dey did return,
dey ain't said nuttin' about it.   25

He was a man dat stood well in society.
He was a member of several lodges.
He was a member in good standing
of Obadiah Lodge number 16-QIXP of XW.
He was also a member of the Chal-deans,   30
de Mis-Carriers Half-Moon Pilgrims,
and de Independent Order of Hen-Roost Disturbers.

De body will be brou:ght here dis mawnin',
an' placed on de left side de church.
Den de congregation, ah,   35
will gadder on de right side de church.
Den de congregation, ah,
will move around from de right side de church
to de left side de church,
and take one last lingering look at de remains   40
while Sister Penny will play dat beautiful hymn,
"All Coons Look Alike t' Me."

Now, a great many people might inquire
what did 'e die of?
Dere's been a great deal o' discussion in dis community   45
about how did 'e die.
I am pleased to state, my brothers an' sisters,
dat 'e died in a glorious manner.
He was shot in de back last Thursday mawnin'

at fo' 'clock g.m.,   50
as 'e was gwine over Miz Grady's back yard fence
wit' six chickens and one duck.

The title, "Colored Funeral," together with the spoken announcement at the
beginning of the recording, orients us to the performance we are about to hear.
The announcement, "Imitation of an old-time colored preacher down South
buryin' one of de brothers, by George Graham," offers interesting contextualizing information. "Old-time" suggests that the preaching style will be anchored
in the past, old-fashioned, somewhat anachronistic. Recall that James Weldon
Johnson consistently used the same adjective to describe the preachers that
inspired *God's Trombones*. "Colored" identifies the style as African American;
taken together with "imitation," it evokes blackface minstrel, vaudeville, and
medicine show performances. "Imitation," as a frame, suggests iconicity but
without the felicity conditions that would confer upon Graham's representation the full performative efficacy of a real sermon. And finally, "down South"
gives the performance a regional grounding: the South as primarily rural,
largely backward, the symbolic heartland of traditionalized Black culture.
Graham's introductory announcement, in other words, encapsulates a chronotope, a representational framework by which expressive genres coordinate
characteristic images of personhood, modes of expression (or voices), and
forms of action in temporal and spatial contexts, establishing the spatial and
temporal coordinates of the blackface sermon as a genre and identifying the
characterological figure who represents its central actor (Agha 2007 b, 177;
Bakhtin 1981, 84–85).[1]

So, with this richness of contextualization cues, how does the recorded
performance actually sound? The sermon starts off in a notably conventional
key, building on recognizable components of the sermon genre. It opens with
a formulaic salutation to the congregation, "my dear beloved brothers and
sisters," echoing scriptural models (e.g., 1 Corinthians 15:58, James 1:19, Philip-
pians 4:1, etc.) and characteristic of American Baptist and Methodist sermons
of the period. The parson then proceeds with a metadiscursive introduction to
his spiritual "text," declaring his intention to broach the authoritative framing
text of the service to come, and then goes on to cite the text itself: "I want to
say one thing dis mawnin', dat in de midst of life, we are in death." Reflexive
phrases like "I want to say" are among the most common formulaic elements
in African American preaching (Rosenberg 1988, 79). This "text," thematically
appropriate to a funeral sermon, is drawn from the Burial of the Dead section
of the Book of Common Prayer, used not only by Episcopalians but also by
Methodists.

Having set out his text, the preacher goes on to apply it to the life circumstances of his congregation, again a conventional step in the development of a homiletic sermon. He couches his delivery in an appropriately churchly register, including a nicely parallel construction with scriptural resonances: "You can see it in de mountains and in de valleys" (cf. Joshua 12:8). Then another formulaic salutation, followed by an exhortation to the congregation to "prepare for dat mighty time t'come," perhaps the life crisis point of their own deaths, perhaps Judgment Day. Either way, the preacher is pretty much on track, and it sounds right—couched in the scriptural register of the prophets, who are always going on about some day or other to come (e.g., Isaiah 42:23, Ezekiel 7:7 and 12, etc.). The appropriate thematic focus for the sermon once established, the preacher arrives, in line 18 and the following lines, at the specific business at hand: memorializing one of their members, newly deceased.

At this point, I'd like to pause and take stock. What I have identified thus far is a series of functional and thematic aspects of the recorded performance that are conventions of the funeral sermon as a genre. It is all conspicuously condensed, subject to the time constraints of the phonograph recording, but the proper slots are appropriately filled. We might ask, then, what it is about the text that marks it as African American. What about dialect? In point of fact, there's less there than one might expect. In terms of phonology, there's some /d/ for /ð/ substitution, as in "dis" and "dat" for "this" and "that," and deletion of postvocalic /r/ and dropped /g/ in /[mɔnɪn]/. That's about it. These are features of African American nonstandard but far from distinctive of that dialect alone (Green 2002, 117–119). As for grammar, it is entirely standard, at least up to line 21, where "am" for the third person singular form of "be" kicks in. Note that in line 8, Graham uses the standard "is." What seems to be happening, then, is that the use of "am" in lines 21–22, coupled with the broad "no mo" and with the aspectual marker "done," in "am done," is meant to signal a code shift, a kind of downward breakthrough of the "old-time," "down South," African American dialect. "Am" for the third person singular was perhaps the most conspicuous grammatical marker of minstrel dialect, though there appear to be questions about how current it was in actual speech (Green 2002, 177, 203; Hurston 2022, 64). "Am done," for its part, does not occur in Black English vernacular (BEV). These forms are conspicuous enough, I would suggest, to change the key of the sermon. The rekeying effect of the dialect downshift is redoubled by the comic name of the deceased, Flatback Jackson, heard for the first time at this point. Flatback, of course, is a ridiculous name, suggesting laziness, out-of-it prostration, and mirroring the supine state of the deceased. The vowel harmony and rhyming of /flætbæk/ and /dʒæk/ adds a further flavor of speech play to the name of the deceased. The low, comical name works in tandem with the dialect shift in effecting the rekeying. This is the point at which the performance turns into burlesque.

Nevertheless, what makes this most strongly an imitation of an African American sermon lies elsewhere. Anyone who listens to the recording will be struck most strongly by the poetic organization and style of delivery of the text, the most conspicuous features of the performance and its most distinctively African American elements, as Johnson anticipated. Inspection of that portion of the performance we have examined reveals it to be segmented into four verses, defined by the initial particles "Now," "Yes," "Ah," and "Now" again. Each of the verses is further defined by falling intonation at the end, sentence completion, topical completion, and—for the third verse—finalization of a parallel construction. The space between verses is further marked by relatively longer breath pauses than those found elsewhere in the text. The individual verses are made up of four or five lines, defined by breath pauses and syntactic structures (phrases or clauses). The cadenced intonational structure of the sermon is especially foregrounded. In general, the tonal range of each line and the text as a whole is very narrow, making for a largely monotonic chant but with occasional accentuated line-internal jumps of a major third and a falling tone in sentence- and verse-final position. The raised tones give the impression of moving from the prevailing chant toward song. A clear example occurs in lines 13–16:

A:::h, my dear beloved brothers and sisters,
prepare for dat mighty time
                  _ t'come.
    /_a:::ll
I want you   to prepare   15
for dat mighty time
                _ t'come.

The poetic and prosodic features I have just identified—all consistent, I might say, with Johnson's own—are core elements of what has been identified by practitioners and scholars as the "heightened" or "elevated" style in African American ritual discourse.[2] The measured, cadenced phrasing, with frequent use of grammatical parallelism and grouping into verses, and the narrowed tonal range, tending toward monotonic chanting, are especially prominent. To look ahead a bit in the transcript, the line-final exhalation "ah!" in lines 35–37 is also a characterizing feature of the elevated style. Taken together, these features are indexes in ritual discourse of spiritual inspiration and divine empowerment (Hinson 2000, 70). As Glenn Hinson (2000, 71) describes this indexical relationship, it "finds its most telling confirmation in sermons, where the...heightened style often emerges after the point of 'elevation,' when preachers are said to start receiving ideas and words from on high." Some interpreters identify the elevated style as a form of spirit possession, in which the preacher simply animates a message authored by the Holy Spirit. Others, however, view the words delivered in the elevated style as originating with the preacher but with their affecting power and heightened capacity to move the hearer supercharged by divine agency (Hinson 2000, 281). The elevated style is a discourse register, a conventionalized constellation of performable features that is the characterizing expressive correlate—the voice—of the old-time African American preacher as a characterological figure and the African American as a genre.

To be sure, there is no suggestion that Graham is in a state of spiritual elevation in delivering his funeral sermon. The imitation frame requires only the replication of generic and stylistic patterns that, under the appropriate felicity conditions, would index spiritual elevation on the part of the speaker being imitated. The rekeying effects of the "am no mo" dialect shift and the identification of the comically named Flatback Jackson, I would suggest, call strongly into question whether the old-style African American preacher Graham is animating is in a state of elevation himself. Graham intends them to be debasing rather than elevating. Here, again, is the core of the burlesque effect.

From this point to the end of the recording, the frame oscillates between morally serious and debased. The elevated style, interestingly, continues throughout, sustained by the chanting intonation, grammatical parallelism, and the voiced exhalations at the end of lines 35 and 37. The verse immediately following the "am no mo" dialect shift maintains the nonstandard/standard grammatical contrast, but after that, the grammar is all standard. The burlesque effect in the remainder of the recording depends on thematic contrasts: the eulogy of the deceased as standing well in society reveals him to be a member of four ridiculously named fraternal lodges, including one that exploits the racist minstrel trope of African American men as chicken thieves; the viewing of the deceased culminates in the playing of the popular "coon song," "All Coons Look Alike to Me"; the "glorious manner" of Mr. Jackson's death is shockingly revealed to have been a shot in the back in the act of stealing "six chickens and one duck," playing again on the poultry thief stereotype; and so on. The recording ends abruptly, without the closure appropriate to the sermon genre, either in its straightforward guise (perhaps a call on the congregation to sing a hymn) or its burlesque minstrel guise (for example, the taking of a collection). Graham reached the tight time limits of the recording before he reached the generic limits of the sermon. So, with the account of Flatback Jackson being killed by a shot in the back, "Colored Funeral" comes to a close.

What does this recorded performance convey to the listener? I want to defer drawing any significant conclusions until we have had a chance to hear additional examples, but I do want to mark a couple of points in a preliminary way. First, there can be no question that "Colored Funeral" is heavily racist. It promulgates the stereotype of African Americans as chicken thieves, and it seeks humor in the extreme violence of shooting a Black person in the back. Moreover, it portrays the preacher as unable to sustain a high discursive tone in his sermon, breaking down at times into broad, nonstandard dialect. Nevertheless, I would suggest that Graham recognized, just as Johnson did, that the language of the "old-time" African American preacher "was not prose but poetry," a virtuosic achievement. I come back to these matters in the conclusion of the chapter.

PEERLESS QUARTET, "NEW PARSON AT DARKTOWN CHURCH"

The representation that I want to consider next is entitled "New Parson at Darktown Church," recorded in 1908 by the Peerless Quartet. "Darktown," of course, was widespread in the popular culture of the day as a racist term for African American neighborhoods and as a context for racist depictions of African American community life. The Peerless Quartet, known primarily as a singing group, underwent many changes of personnel over its life course but enjoyed enormous popular success throughout the acoustic era, that is, through the late 1920s (Brooks 2020, 58–63; Gracyk 2000, 267–272).

The Peerless Quartet had a broad repertoire of popular songs and employed a number of presentational formats in their recordings. The songs were essentially vehicles for the display of four-part harmony, not parody. One of the formats the quartet used relatively often was the dramatic representation of African American ceremonial events—weddings, church services, lodge meetings—as a frame for the performance of purportedly Black songs: some genuinely so, others drawn from the minstrel stage. "New Parson at Darktown Church" employs this format, with the church service providing two slots for songs: one at the opening of the service, the other at the close.[3] The sketch is in two principal parts. In the first, the departing parson opens the service, announces the first song for the congregation, and then introduces the new parson, Brother Luther Wilberforce. Brother Wilberforce then takes over, and the remainder of the service consists of his sermon and the closing hymn.

Peerless Quartet, "New Parson at Darktown Church" (Victor B5081-1; https://purl.dlib.indiana.edu/iudl/media/r074855b1q), Feb. 14, 1908

Departing Parson:
While the collection am being took,
the congregation will rise and sing "In the Sweet Bye and Bye."
Now, as your parson has recovered from the affliction of brown-chitis,
it am not necessary to put any more cough drops in the contribution box.

Choir:
[Sings "In the Sweet Bye and Bye."]   5

Parson:
Brethren and Sistren,
Dis am de last time I shall be with you as your parson.
I am goin' to prepare a place for you
dat where I am,
dere may you be also.   10
I have been appointed chaplain of the colored wing
of the Tennessee State Prison.
I now introduces your new parson,
Brother Luther Wilberforce of Memphis.

Brother Luther Wilberforce:
Brothers, sisters, congregation,   15
my text am am dere or am dere not a hell.

Congregation:
Course dey's a hell! Course dey's a hell!

Brother Luther Wilberforce:
Ingersoll said that there was no hell.

Voice from Congregation:
Parson, who was Ingersoll?

Brother Luther Wilberforce:
Why, Andy, I'm astonished at your ignorance.   20
Ingersoll was de man what invented de dollar watch.
Now if dere ain't a hell, dere am gwine to be.
De Lord made earth to turn round on its axletree once in twenty-four
hours.

Congregation:
Oh, yes! Amen!

Brother Luther Wilberforce:
And then he filled earth with oil for to grease de axletree,   25

Congregation:
Yes! Dat's right! Dat's what 'e did!

Brother Luther Wilberforce:
and de Standard Oil Company bored de holes in earth to distract de oil,

Congregation:
[Unintelligible]

Brother Luther Wilberforce:
and den dey moved it to Ohio and dey found it dar,

Congregation:
Oh, dey found it dar!   30

Brother Luther Wilberforce:
and den he moved it to Virginia and dey found it dar,

Congregation:
Yeah! Yeah!

Brother Luther Wilberforce:
and den he moved it to Texas and now dey done found it dar.

Congregation:
Amen! Dey did!

Brother Luther Wilberforce:
And now, when it am all gone den what be dar?   35

Congregation:
What, what, what, what, what?

Brother Luther Wilberforce:
Why, de axle run hot,
de world caught fire,
won't dat be hell?

Congregation:
Course, dat'll be hell!   40

Brother Luther Wilberforce:
And den will descend dat golden chariot.

Congregation:
Amen! Amen!

Brother Luther Wilberforce:
Swing low, sweet chariot.

Choir:
[Sings "Swing Low, Sweet Chariot"]
(Choir) Chorus   45
(Solo) I looked over Jordan and what'd I see?

Voice from Congregation:
What you see, brother?

Choir:
[Sings: "A band of angels . . ."]

Leading up to the sermon itself are three humorous bits that set the tone of the performance, two in the first section and one in the second. First, the departing parson cautions the congregation against putting cough drops in the collection box. Ministerial complaints and exhortations about the insufficiencies of the collection are a convention of the mock sermon as a genre, and the substitution of other objects for cold cash is also a stock motif (Gilman 1974, 67–68; Russell 1991, 242). This little routine, then, can be used to mock members of any denomination as tight-fisted and any minister as venal. Here, it is targeted against African Americans. More powerful as an instrument of burlesque is the carnivalesque subversion of one of the most foundational passages in the New Testament (John 14:3), in which Jesus says, "And if I go and prepare a place for you, I will come again, and receive you unto myself; that where I am, *there* ye may be also." Jesus's preparation of a place in his father's house for all humankind, by his sacrifice, becomes the parson's preparation of a place for his parishioners in the state prison, by his acceptance of a new civil-service position. Interestingly, the passage from John might serve quite appropriately as the "text" for a sermon, but here it is gratuitous, inserted solely for the purpose of setting up the inversion that establishes the burlesque key of the performance.

The third ludic routine occurs in Brother Wilberforce's lead-in to his sermon. Having set the theme of the sermon by posing the question of whether there is a hell (not, note, a true scriptural "text" of the kind most suitable for a sermon), the parson cites some authority named Ingersoll as denying the existence of hell. When asked by a congregant, "Who was Ingersoll?," Brother Wilberforce very authoritatively misidentifies him as the inventor of the dollar watch (the Ingersoll Company was the manufacturer of cheap watches) rather than as the renowned political leader and orator Robert G. Ingersoll (1833–1899), an outspoken religious agnostic. Why cite a watchmaker in a sermon? Again, there's no logic; it's simply an opportunity to show up the pastor as lacking the knowledge and authority he claims.

We arrive, then, at the body of Brother Wilberforce's sermon, expanding upon his "text," "am dere or am dere not a hell?" (Here's our minstrel-dialect "am" again.) The development revolves (excuse the pun) around the pastor's paronomastic confusion of the axis of the earth with the axletree of a wagon, motivated not only by the phonological correspondence between the two terms but also by their semantic affinity: both represent a point around which an object turns. By tropic extension, the axletree of the earth, like the axletree of a wagon, requires lubrication; deprived of lubrication it will ignite. At a time when the oil industry was burgeoning and Standard Oil was at the height of its monopolistic dominance, Brother Wilberforce suggests—note the malapropism—that the "distraction" of oil, placed by God within the earth to lubricate its revolutions, will lead ultimately to the overheating and burning of earth's axletree, and that will bring about a hell on earth.

In formal terms, following the preacher's positing of an initial answer to his "text"—"Now if dere ain't a hell, dere am gwine to be" ("gwine" is another emblematic bit of minstrel dialect; see, e.g., Holmberg and Schneider 1986, 31)—the sermon unfolds in a series of parallel constructions of increasing presentational intensity, building toward a climax in which the earth catches fire and the hell on earth is realized. The lines making up the parallel constructions are defined by syntactic structures and breath pauses and the parallel units are marked off by falling intonation and responses from the congregation, acted by other members of the quartet. The first parallel set establishes the axletree / lubricating oil / extraction frame of reference; the second charts the acceleration of the extraction process until the oil supply is exhausted.

> De Lord made earth to turn round on its axletree
> once in twenty-four hours.
> And he filled earth with oil to grease de axletree.
> And de Standard Oil Company bored de holes in earth to distract de oil.
>
> And den dey moved it to Ohio,
> and dey found it dar.
> And den he moved it to Virginia,
> and dey found it dar.
> And den he moved it to Texas,
> and now dey done found it dar.

The concluding section begins with the parson's question, "And now, when it am all gone, / den what be dar?," which is linked to the end of the preceding section by repetition of the line-final "dar." When the congregation asks him, "What?," Brother Wilberforce concludes with a final parallel set and an answer to the question posed by his "text":

> Why, de axle run hot,
> de world caught fire.
> Won't dat be hell?

And into that apocalyptic inferno of hell on earth descends the golden chariot of the cherubims (I Chronicles 28:18)—to save the righteous, perhaps?—and the service ends with the congregation, animated by the quartet, singing the classic African American hymn "Swing Low Sweet Chariot."

As in "Colored Funeral," Brother Wilberforce's sermon is rendered in a variant of the elevated style. I should perhaps note here that there is considerable latitude for stylistic variation on the part of individual preachers; what matters is the contrast between ordinary talk and heightened preaching. Like our earlier example, Brother Wilberforce's sermon is chanted, marked by a severe narrowing of the tonal range and a falling intonation at the ends of lines. This sermon is also marked by an incremental raising of the pitch of successive parallel lines and a pronounced shortening of the lines as the sermon reaches a climax, two further characteristics commonly found in elevated preaching. Perhaps most importantly, the lines delivered by Brother Wilberforce are also defined as turn-transition points, open to responses from the congregation that index their own heightened spiritual state. George Graham, as an individual performer, did not have others to enact the part of congregation members, whereas the Peerless Quarter had the personnel to build in this important and strongly characteristic feature of the African American sermon. "New Parson at Darktown Church" thus expands our inventory of devices employed in the burlesque sermons to represent the elevated style. Brother Wilberforce may have his authority compromised by mistaking the watchmaker Ingersoll for the agnostic orator Ingersoll; he may wind himself up in a spurious metaphor revolving around the earth's axletree; but he is nevertheless able to sustain the core metaphor with formal rigor and affecting power toward a spiritually heightened finale. The new parson's sermon is no mean poetic achievement.

RALPH BINGHAM, "BROTHER JONES' SERMON"

My third example, issued in 1918, is "Brother Jones' Sermon," by Ralph Bingham, a performance routine popular enough to have become a household performance piece, according to a colleague who recalled versions of it performed by family members in the late 1940s. Bingham was the consummate,

multitalented lyceum entertainer.[4] He began his platform career in 1876, at the
ripe age of six, as "The Boy Orator of America," reciting poetry and speeches
and playing the violin. As a mature entertainer, Bingham was billed as "per-
sonator, humorist, violinist, vocalist, raconteur," according to one of his publicity brochures, and that flier neglects to mention that he also played the piano.
The core of his professional identity, though, was humor: monologues featuring rural life and sports, storytelling (including some traditional folktales),
and—most especially—dialect, listed in the record catalogs as "colored" or
"Negro dialect."

**Ralph Bingham, "Brother Jones' Sermon" (Victor 18587-B, mx. 21405;
https://purl.dlib.indiana.edu/iudl/media/f16c08qb9x), Jan. 4, 1918**

Brother Jones:
My brethren,
I take my text this evenin'
from the fo'teenth verse o' the fo'teenth chapter
accordin' to the Gospel of Etymology.
"And de Lord cured the multitude   5
of divers diseases."
My beloved brethren and sisters,
does you all get dem words?
Does you all corroborate dar dimension?
Does you all qualify dar intrusion?   10
Hmm?
Does you all specify de impo'tance of 'em?

Voice from Congregation:
What'd he say?

Brother Jones:
Let me 'lucidate de words again, my brethren.
"And de Lord cured de multitudes   15
of divers diseases."
Does you all notice dat it don't say nothin' about the *plu*-ralisis,
or de *phew*-monia?
No, suh!
It don't say nothin' 'bout yella ja'ndice or yella fever.   20
No, suh!
It don't say nothin' 'bout de *pen*-deceetis
or de spiral mcginnis.
No, suh!
What do de good book say?   25

Hmm?!
I asks you what do de words transmogrify?
Dey say:
"And de Lord cured de multitudes
of divers diseases."   30
Does you notice, my brethren,
dat dey ain't nothin' 'bout de Lord wastin' valuable time
curin' de multitudes of little ol' common miseries,
like *mule*-aria,
or *ty*-phoid,   35
or chills an' fever?
It sure don't.
What do de words of wisdom equivocate or expectorate?
Hmmm?
Hyar dem again brethren and sisters of de faith.   40
Hyar dem again.
"And de Lord cured de multitudes
of divers diseases."
Oh, my chilluns.
Oh, my lambs of Zion.   45
Oh my pillas and bolsters o' de church.
Brudder Erysipelas Brown,
will you tell dem white boys sittin' back by de stove
dat if dey don't stop laughin' and behavin' injurin' of de sermon,
dat we'll send for the town constibule and outen 'em.   50
Oh, my lambs o' love.
What do de text mean?
Oh, my brethren and sistren,
if dey gets the janders,
any little ol' two by four phizzican can cure ya.   55
If ya gets the *phew*-monia,
any little ol' pillbox can cure ya.
If ya gets the *neu*-ralgy,
any little ol' sawbones can cure ya.
But oh, my brethren!   60
Oh, my brethren.
If ya once gets the divers,
mm mm!
You're gone!
You're de gamecock in de pit.   65
Nobody but de Lord hisself can cure da divers.
"And de Lord cured de multitudes

of divers diseases."
Nobody but de Lord, my lambs, can cure de divers diseases.
De worst disease is what . . .   70
Ah-choo!
Ah-choo!
Ah-choo!
We will now take up a collecti—
Ah-choo! choo!   75
And disband de meetin'.
For some o' dat low-down, cheap white trash
has done throwed red pepper on de flo'.
Aaaah-choo!

By now, we're familiar with the core characteristics of the elevated style.
Bingham employs a veritable anthology of devices in his representation: chant,
repetition, parallelism, formulaic salutations and other phatic gestures to the
congregation, and so on. I won't analyze the formal organization of Brother
Jones's sermon in detail—it's too complex for the space I have available to me.
Let's just have a quick look at one brief section (lines 38–64):

What do de words of wisdom equivocate or expectorate?
Hmmm?
Hyar dem again brethren and sisters of de faith.   40
Hyar dem again.
"And de Lord cured de multitudes
of divers diseases."
Oh, my chilluns.
Oh, my lambs of Zion.   45
Oh my pillas and bolsters o' de church.
Brudder Erysipelas Brown,
will you tell dem white boys sittin' back by de stove
dat if dey don't stop laughin' and behavin' injurin' of de sermon,
dat we'll send for the town constibule and outen 'em.   50
Oh, my lambs o' love.
What do de text mean?
Oh, my brethren and sistren,
if dey gets the ja'nders,
any little ol' two by four phizzican can cure ya.   55
If ya gets the *phew*-monia,
any little ol' pillbox can cure ya.
If ya gets the *neu*-ralgy,
any little ol' sawbones can cure ya.

But oh, my brethren!   60
Oh, my brethren.
If ya once gets the divers,
mm mm!
You're gone!

The passage displays the marked intonation patterns of heightened preaching, verging here on song, the measured cadences, the grammatical parallelism, the phatic formulas that we might expect. I choose it, though, because it also illustrates clearly the contrast between the elevated style and ordinary talk, when Brother Jones breaks frame at line 47 to direct a member of the congregation to quiet the White intruders to the service and then switches back into sermon style at line 51. The unmarked, conversational passage features longer, unmeasured utterances between breath pauses, a raspier timbre, no parallelism or repetition.

This extract serves well as an example in one other respect: the speech play in line 46, "Oh, my pillas and bolsters o' de church." "Pillars of the church," of course is a cliché, a commonplace, and "bolster" may also mean "to shore up," "support." Dialect renders "pillars" and "pillows" homophonous; the latter term then turns away from "pillars" to pair with "bolsters" in the same, stuffed-cushion semantic field. That bit of speech play, of course, is but one of many instances in Bingham's performance. Indeed, one might argue that the entire routine turns on speech play and complementary forms of metalinguistic and metadiscursive reflexivity.

The central trope, clearly, consists in Brother Jones's misconstrual of his scriptural "text," taking the "divers diseases" of Matthew 4:24 and Mark 1:34 to be a single, specific, deadly malady, susceptible only to the Lord's miraculous curing power. As the trope unfolds, then, in a profusion of parallel constructions, the preacher adduces an entire inventory of diseases that are merely "little ol' common miseries" that can be cured by routine medical treatments: "any little ol' two by four phizzican," "pillbox," or ordinary "sawbones." As Brother Jones reels off these "common miseries," he makes a paronomastic mess of most of them, in portmanteau mergers like "pluralysis," "mule-aria," "phew-monia," or in puns like "spiral mcginnis" (for "viral meningitis," I assume), not to mention simple but nonstandard vernacular pronunciations like "neur-algy" and "janders."

In ironic tension with all these paronomastic misfires is a steady counterpoint of metalinguistic and metadiscursive monitoring on Brother Jones's part to ensure that the members of his congregation comprehend the meaning and significance of his scriptural text. This reflexive strain in the sermon is adumbrated by the very attribution of the text itself to the Gospel of Etymology. Brother Jones is in love with fancy words; Black and White observers alike identify this penchant as a common characteristic of African American preachers (J. Johnson 1976, 9; Mahar 1985, 263; Washington 1909, 2:284–285). For burlesque sermons, this predilection and the susceptibility of uneducated preachers (again Black or White; see McCurdy 1969, 165) to mispronounce, misuse, or otherwise distort words because of their morphological complexity or misleading spelling is a ready comic resource. Immediately after citing his text, the preacher makes his first comprehension check, in a burst of malapropisms:

> My beloved brethren and sisters,
> does you all get dem words?
> Does you all corroborate dar dimension?
> Does you all qualify dar intrusion?   10
> Hmm?
> Does you all specify de impo'tance of 'em?

And throughout the remainder of the sermon, Brother Jones keeps checking in the same vein (as in "what do de text mean?" in our earlier extract), asking rhetorical questions, pointing out key features of the text, and commenting on his own efforts to make things clear—though his malapropisms offer much more amusement than clarity.

Beyond the centrality of speech play and metadiscursive reflexivity in "Brother Jones' Sermon," what is especially interesting about this performance is the depiction of White disruption of a Black church service and the preacher's resistance to the intruders. He may be a figure of ridicule, in his misconstrual of his biblical text and the mess he makes of medical terminology, but he has the moral strength to tell off a group of White rowdies. And, like Graham's "old-time colored preacher" and the Peerless Quartet's Brother Wilberforce, Brother Jones is clearly a verbal artist, an oral poet of virtuosic ability.

BERT WILLIAMS, ELDER EATMORE'S SERMONS

My last two examples, which I'll consider together, represent something of a departure from the other texts we have examined. Both were recorded in 1919 by Bert Williams, one of the most celebrated vaudeville performers of the day and a highly popular recording star as well (Brooks 2004a, 10–148; Forbes 2008; Reed 2001; Rowland 1923; E. Smith 1992). Significantly, Williams was Black, one of the many African American performers who appeared in the burned-cork makeup and stereotyped costume of the blackface minstrel tradition. I won't go into this complex phenomenon—many others have done so—but Williams was unique even in that company. Born in Nassau, he came to the US with his family at the age of ten and grew up in Riverside, California. He aspired to be a civil engineer but drifted into vaudeville in San Francisco and rose to stardom as a singer and comedian. He was the first Black member of the Ziegfield Follies, attracted an enthusiastic following of fans, both White and Black, and was the best-selling Black recording artist of the day, by far. Significantly, he was lauded—both for his talent and his success—by African American cultural leaders ranging from Booker T. Washington to W. E. B. Du Bois (Brooks 2004a, 123; Forbes 2008, 95–96, 321; E. Smith 1992, 145).

It is important to establish that for Williams, blacking up and taking on African American dialect were acts of Othering more akin to the transformations effected by White blackface performers than by Black ones. Fair-skinned and relatively well educated, he identified himself as Bahamian; the African American dialect of the US mainland was not a part of his native repertoire. While he accepted—he certainly could not escape—his identification as a Black man in the US, he occupied a position at a greater remove from the African American types he enacted than his fellow Black performers were able to achieve. More on this later, after we consider his burlesque sermons.

Let us first consider "Elder Eatmore's Sermon on Throwing Stones," in my view, the more interesting of the two 1919 recordings. Elder is the term for preachers in the "holiness" or "sanctified" sects that emphasized adherence to strict moral standards and avoidance of carnal activities (Baer and Singer 1997, 267).The title of the recording aligns it to the burlesque sermons from the minstrel tradition, and it bears certain features of the genre, but the actual performance is more a comic character sketch than a parody of the sermon itself.

Bert Williams, "Elder Eatmore's Sermon on Throwing Stones" (Columbia A6141, https://purl.dlib.indiana.edu/iudl/media /t44p58w953), June 27, 1919.

The Elder is in bad humor
dis mawnin'.
I take my text from de 'leventh of d'Ecclesiastides
"He dat is of mostly widout sin,
let 'im throw de fust rock."   5
What it says is,
"Let him cast de fust stone."
But I ain't takin' no chances on you all misunderstandin' me.
For twenty years y'all be thowin' rocks at one another,
but you wasn't satisfied.   10
You had to commence thowin' 'em at me.

Voice: Uh uh!

Fig. 2.1. Publicity photograph, Bert Williams in blackface, 1921. Library of Congress Prints and Photographs Division, Washington, DC.

Fig. 2.2. Publicity photograph, Bert Williams portrait, 1922. Library of Congress
Prints and Photographs Division, Washington, DC.

Uh uh nothin'!
But I ain' gon' warn y'all no mo'.
But in de language of dat great prophet Henry Shakespeare,   15
"Watch yo' step,
w:::atch yo' step."
What did Nicodemus say?
I says, what did Nicodemus say?
He said, "wash me an' I shall be whiter den snow."   20

Voice: A:::men!

Mhmm hmm.
Dat's all right too.
Dey's a lot o' y'all here dis mawnin'
think you been washed.   25
You ain't even been sponged.

Voice: Hunh!

Nh. T'ain't no use gruntin'.
On night befo' last Thanksgivin',
I think it was long about midnight,   30
certain brother—he's sittin' right here out there—
dis brother was comin' down de road,
totin' a bag,
an' he sees another brother,
totin' another bag,   35
an' gettin' ovuh a fence.
Bot' bags was occupied.
Now neither one of dese brothers spoke,
but a sound fum de bag
of de brother on de fence   40
indicated dat he had secured de main article fo' his Thanksgivin' dinner.

Voice: Oooh!

Uh huh!
Now,
on dat other certain brother had   45
in his sack
a member of de same family,
but it wasn't
'zackly de kind of a bird
gen'ly used fo' Thanksgivin' dinner.   50

Voice: Mmm hmm.

Mmm hmm.
An' dis filled dat certain brother's heart
so full of jealousy
an' malice   55
dat he goes straight home
an' tells 'is wife what 'e'd seen,
an' dat does settle it.

Voice: Ain't it de truth!

Unnh.   60
You know it's de truth!
You know it's de truth!
His wife tells
her sistuh,
her sistuh tells her friends,   65
an' de fust thing I knows
ever' membuh of dis congregation here
is whisperin' around
dat I,
me,   70
me,
hah!
had stole a turkey.

Voice: [Whistles: phew!]

Now, dey ain' no use phewin'.   75
In de future,
any of you all dat thows rocks at me,
I'm gon' thow 'em back at you.
An' when I
start   80
to thowin',
friends, I shall miss nobody.
Dere is silence.
Now dat certain brother that I've been talkin' 'bout
will kindly lead us in prayer.   85

Voice: There now.

De scriptures say
dat who de gods would last destroy,
dey fust makes mad.

EE: Oh, no now!   90

Voice: An' Elder Eatmo' sho is actin' crazy!

EE: Now there, there, there, there! There!

Voice: On las' Thursday night,
who was it
dat I had to almost tote home?   95

EE: No, no,
wait a minute!

Voice: Dat I almost had to carry home bodily.   //EE: Wait a minute!
    Don't answer now! [?]
    Now look here!   100
                                        [Unintelligible]

Voice: Yes,
he was so full of applejack.
Hnnn.
An' who was it,
                                            // EE: Now here!   105
who was it dat steady stole de lodge's money?   Dat's enough!
I said, who was it Lo:::rd   I say, dat's enough!
dat stole de lodge's money?
An' lost it playin' five up down at Sister Mamie Crawford's?
Who . . .   110

EE: Doxology,
doxology,
doxology.
[Organ music]
Use all de doors.
Use all de doors.   115
Use all de doors.
We are all leaving now.
All leaving.

The elder opens with a general expression of ill humor and then devotes his
entire disquisition to rebuking those who have put him in his bad mood and
threatening to respond in kind. His "text," misquoted and misattributed, as we
have come to expect in burlesque sermons (it's actually from John 8:7), makes
clear the source of his disgruntlement: he is the target of malicious accusations
by members of his congregation. In response, he goes on a counteroffensive,
warning his accusers to back off, with further spurious, misattributed "texts,"
from the "great prophet Henry Shakespeare" and Nicodemus, the sympathetic

Pharisee mentioned in the book of John. Notwithstanding its scriptural source (Psalms 51:7), "Wash me and I shall be whiter than snow" is a stock bit from the minstrel tradition, a nasty reminder, packaged as a laugh line, that Black people can never be White and thus, perhaps, can never be spiritually pure. Certainly, that's what he claims of his unsponged congregants. Elder Eatmore goes on to single out the specific parishioner he holds responsible for spreading malicious rumors about him but winds up revealing in the process that he has indeed been out on a poultry-stealing expedition during which he encountered this "certain other brother" likewise engaged—the old racist minstrel trope of Black men as chicken thieves.[5] In the elder's mind, it was that "other brother's" jealousy, provoked by the preacher's having bagged a turkey while he had only a chicken, that led to the backbiting accusations against him. The other brother tells his wife what he has seen, his wife tells her sister, her sister tells her friends, and the gossip against Elder Eatmore spreads quickly through the congregation. The elder is at pains not to name names, as if he were above the kind of malicious gossip that is directed against him, but he still threatens retaliation: those who malign him will be the target of his own barrage of counteraccusations.

Although the elder's rant is framed by a biblical text, there is nothing spiritual about his message. The only hint of the heightened style is in a few parallel lines ("watch yo' step / wa:::tch yo' step"), inspired by a mean spirit, not the Holy Spirit. And the responses from the congregation are not the usual enthusiastic ratifications of the preacher's inspired words but dismissive ("Hunh!"), shocked ("Oooh!"), or incredulous ("Phew!") reactions to his peevish accusations and threats.

The tenor of the performance shifts, though, when Elder Eatmore, perhaps thinking that he has cowed the nameless parishioner who has accused him, calls on him to lead the congregation in prayer. The "other brother" leaps immediately into the heightened style, but not in prayer. Rather, in a vehement counterattack on Elder Eatmore, he accuses the preacher of further moral lapses— drunkenness, embezzlement, gambling, frequenting low places—painting him as the antithesis of what a holiness preacher should be. The elder tries to interrupt him but is powerless to stop the harangue—the "other brother" is too forceful and caught up in his tirade. The elder's only recourse is to bring the service to a hasty end and evacuate the church.

The other "sermon" recorded by Bert Williams, "Elder Eatmore's Sermon on Generosity," is consistent with the first in its depiction of his character and his relations with his congregation, though not as rich a performance (https://purl .dlib.indiana.edu/iudl/media/c38653dv52). The elder takes his text this time from the Book of Caesar: "the Lord loveth a cheerful giver." There is no Book of Caesar, of course. The text comes from 2 Corinthians 9:7, but the misattribution indexes the famous passage in Mark 12:17, "Render unto Caesar the things that are Caesar's, and to God the things that are God's." Either way, Elder Eatmore wants his share of the proceeds. The scriptural theme notwithstanding, however, the ensuing "sermon" is a very worldly and self-interested exhortation for the congregation to cough up more money for the elder's support, because they are "way back in my salary" and "I need! I need!" It won't do for Elder Eatmore to have to eat less. The castigation of the congregation for its tight-fistedness is a venerable theme in parodic sermons (recall the opening of "New Parson at Darktown Church"); Gilman cites a medieval example of a "poor country preacher, striving to obtain the necessary financial support from his congregation" (Gilman 1974, 67; cf. Russell 1991, 242, 248–249). Interestingly, though, the theme seems to have figured sufficiently strongly in actual Black preaching for Pipes (1945, 17) to have noted it in his early (1942) field study of Black sermons in Macon County, Georgia.

Elder Eatmore's straitened situation is exacerbated by the increasingly "scientifical" security measures that make it "harder for all of us" to gain access to smokehouses and henhouses. Accordingly, he exhorts his parishioners with an extended series of scriptural allusions (for example to "the bread of life"), proverbs ("the Lord helps them that helps theirself"), and hymn titles and lyrics ("We shall build our mansions in the sky," "We shall reap our joys in the bye and bye") to provide for his material needs. All the high-sounding quotations and allusions he reels off might well lend themselves to spiritually and morally uplifting ends, but here they are solely in the interest of persuading the elder's tight-fisted parishioners to open their purses and provide more liberally for his support. Taking no chances, though, as passing the plate has proven unreliable, he calls the congregation to file past him one by one and drop their money on the table before him. The message of the performance is ambiguous. It may be heard as grasping and self-interested extortion by a venal preacher or as an indictment of stingy congregations, but either way, there is little spirituality or religious inspiration in "Elder Eatmore's Sermon on Generosity," only a conspicuously worldly appeal for cash.

As I suggested earlier, Bert Williams's Elder Eatmore recordings are not so much burlesque sermons as they are character sketches of the flawed preacher. Unlike the examples we have considered by White blackface performers, the poetics of the Black sermon and the artistic virtuosity of the Black preacher are not foregrounded in these performances. But what kind of character does Williams enact? On the one hand, I would propose, Elder Eatmore is aligned to the long comic tradition of the morally compromised preacher: self-absorbed, venal, vindictive, hypocritical, doctrinally ignorant, and—if his name is any indication—greedy. But the elder is also more than simply a blackface version of a venerable comic type. Williams himself is explicit about his focus on character. When asked by an interviewer what aspect of his work interested him most, Williams replied, "Character," and went on to state, "I try to portray the shiftless darky to the fullest extent.... There is nothing about this fellow I don't know. I have studied him" (Rowland 1923, 94). While Williams was at pains to portray the folk wisdom of his "shiftless darky" characters, his very use of the term makes clear his willingness to sustain the negative aspects of stereotype as well, and I would suggest that Elder Eatmore foregrounds that tendency. I have much more to say about this in my conclusion.

CONCLUSION

For White performers, the "old-time colored preacher down South," to use George Graham's characterization, was a variant on the nostalgic construction of the "old-time Negro," invented to valorize a mythic past in which Black people were unthreatening, ignorant, yet expressively attractive and potentially entertaining.[6] The old-time preacher and his sermons served the White interests well, simultaneously showcasing how even the moral and intellectual leaders of the Black community were ignorant in their use of scriptural sources, illogical in the development of homiletic themes, prone to linguistic misfires in their use of cultivated liturgical registers, but capable, withal, of impressive, virtuosic performance. Parody was an especially effective resource for White entertainer's purposes, insofar as it allowed for manipulation of the intertextual links and gaps between their own mock sermons and the African American source genre. White performers in blackface could undermine the authority and spiritual power of the sermon by rendering the thematic features of the sermon ridiculous and marking the linguistic—grammatical, phonological, and lexical—aspects of the delivery as incompetent but preserving and even foregrounding the expressive virtuosity of the delivery as a showcase for their own skill as performers.

What is most striking about the White parodies of Black sermons, I would argue, is that notwithstanding the incompetencies and infelicities with which he is burdened by the mocking portrayals of the White performers, the preacher is represented as an oral poet, with an impressive mastery of the verbal art of preaching. Even within the truncated format of the recording, the preachers all launch their sermons in a form appropriate to the genre, and while Graham's preacher is cut off by the time limit of the record and Brother Jones by the delinquent White boys who disrupt his service, the fault is not theirs, and Brother Wilberforce at least carries his sermon to appropriate completion. Far more importantly, however, all three preachers deliver their sermons in a recognizable and highly credible variant of the elevated style, clearly marked by multiple poetic devices and patterning principles: initial particles, intonation contours, breath pauses, emphatic exhalations, congregational responses, grammatical parallelism, organization into verses. None of the published minstrel-show sermons I have seen come close to this degree of poetic complexity. Indeed, I believe strongly that for all that they draw on minstrel conventions, all of our performers must have had direct, firsthand familiarity with Black preaching to have represented it in such detail. These White blackface performers were closer observers than many critics realize. At least regarding sermons, they displayed impressive ethnopoetic understanding, a quarter century before James Weldon Johnson's effort to recuperate and revalue the Black sermon. Moreover, they all recognized in the style of the African American preacher an artistic accomplishment worthy of their own performance skills, an appropriate vehicle for their own displays of virtuosity.

For Black performers, artists, and intellectuals, the situation was far more complicated. Among intellectual leaders, especially, there was a profound preoccupation from the mid-1890s through the first several decades of the twentieth century with the need to shed the burdens of an oppressive past and the persistent shadow of debilitating stereotypes that stood as impediments to social, economic, political, and cultural advancement. Increasingly useful as summarizing tropes in the effort to recreate African American society and culture, from the mid-1890s onward, was the contrast between the "old(-time)" and the "New Negro," inscribed in such influential works as Booker T. Washington, Fannie Barrier Williams, and N. B. Wood's *A New Negro for a New Century*, published in 1900, and Alain Locke's cultural manifesto, *The New Negro*, published in 1925 and soon established as the charter of the Harlem Renaissance.

In the eyes of many, Bert Williams was the shining epitome of the New Negro. He was educated, economically successful, Episcopalian, and enthusiastically admired by Black and White audiences alike. W. E. B. Du Bois and Booker T. Washington, who were vehemently opposed in their conceptions and strategies for the realization of the New Negro, nevertheless agreed that Bert Williams exemplified the best of the race. While Du Bois praised Williams's great skill as a performer (DuBois 1924, 310), Washington was more explicit and effusive about his potential contribution to the cause of Black advancement: "Bert Williams is a tremendous asset of the Negro race. He is an asset because he has succeeded in actually doing something, and because he has succeeded, the fact of his success helps the Negro many times more than he could help the Negro by merely contenting himself to whine and complain about racial difficulties and racial discriminations" (Brooks 2004a, 123). Moreover, Williams aligned himself with Washington's ideology and program. Responding to Washington's praise, Williams suggested that "the negro actor will . . . take rank with the negro teacher in the negro school. Booker Washington will then have strong allies in his work of elevating the social standard of the black man" (E. Smith 1992, 110).

While all who were engaged in this effort of symbolic construction agreed on the need to overcome the debilitating effects of racial stereotypes, there was considerable debate about what elements of the Black cultural experience might be worthy of retention, recuperation, and, perhaps, re-creation, in the effort to create the New Negro. African American religion, understandably, occupied an important place in these debates. Some intellectual leaders, including prominently W. E. B. Du Bois, insisted on the deep centrality of the Black church—itself a symbolic construction of these debates—to the integrity and spiritual sustenance of African American people throughout their existence in the New World, celebrating the Black preacher for his "singular eloquence" and the moral power of his oratorical skill (Du Bois [1905] 2003, 191). At the same time, however, even Du Bois (1924, 331–332) expressed some ambivalence about the "stirring and wild enthusiasm" and "the wilder spiritual emotionalism of the black man," reaching back to "the unlettered childhood of the race rather than to the thinking adult life of civilization." For those of "the better classes," as Du Bois (2003, 207) called them, who believed that the new way for the New Negro lay in the direction of discipline, education, and cultivated and refined decorum, the unlettered enthusiasm of the old-time holiness preacher was an embarrassment, a corroboration of the stereotype of the Black person as ignorant, unsophisticated, undisciplined, and linguistically incompetent. All the worse when the preacher also embodied the stereotype of the manipulative, immoral schemer, like Elder Eatmore, a discredit to his religious office.

Parody is a vehicle of critique, and Bert Williams's Elder Eatmore sermons aim their critical barbs against aspects of African American "old-time" religion that the champions of the New Negro considered an impediment to Black respectability and advancement. Undermining the stereotype by enacting the stereotype is a tricky business, though, especially insofar as Williams played to enthusiastic White audiences at the same time that he was lauded by Black people. Elite Black audiences could hear Elder Eatmore's sermons as a critique of old-time holiness preachers they wanted to leave behind, but for Whites, the performances may well have perpetuated the very images that had burdened African American people all along.

As I suggested at the beginning of the chapter, however, there were other participants in the effort to redefine African American culture in the early decades of the twentieth century who had different conceptions of the value of "old-time" vernacular religion and how it might be recast to the credit of African American people. James Weldon Johnson, in particular, recognized the affecting power of the "folk sermon" and the poetic skill of the old-time preacher and saw in them a potentially valuable resource for a new "Aframerican" literature. More specifically, he sought to capture the preacher's masterful use of rhythm, intonation, tempo, timbre, parallelism, and the other devices we have examined—in conjunction, of course, with the spiritual resonances of religious themes and symbols—in fashioning a new register for African American poetry to replace the compromised language of "*traditional* Negro dialect" (J. Johnson 1976, 6; italics in the original). The Aframerican poet, he insisted, "needs a form that is freer and larger than dialect, but which will still hold the racial flavor" (J. Johnson 1976, 6). The poetic devices of the traditional sermon, he suggested, might represent the basis of just such a form, separable from "the mutilation of English spelling and pronunciation" (J. Johnson 1976, 6) and able to stand on its own as a poetic register. Thus Bert Williams and James Weldon Johnson represented strongly contrastive stances on the "old-time" preacher and his sermons within the larger cultural movement to construct the New Negro and transcend the old negative stereotypes. The White parodies, while they exploited the same poetic power that Johnson built upon, were nevertheless vehicles for the preservation of a racist past, foils against which the African American intellectuals and artists sought to imagine a new future—and a new chronotope—for their people.

NOTES

1. The notion of chronotope was introduced and most fully developed in Bakhtin 1981. For useful elaboration of the concept in linguistic anthropology, see Agha 2007a, 2007b; Blommaert 2015; Briggs 2007a; Lempert and Perrino 2007.

2. There is an extensive literature on the African American oral sermon. I have used the following works: Davis 1985; Gumperz 1982, 187–196; Hinson 2000; J. Johnson 1976; LaRue 2003; Mitchell 1970; Oliver 1984, 153–155; Pipes (1951) 1992; Pitts 1989; Raboteau 1995; Rosenberg 1988; Wharry 2003.

3. This routine was authored by Cal Stewart although he did not perform it.

4. Information on Bingham is available online at the University of Iowa Digital Library Archives, Traveling Culture: Circuit Chautauqua in the Twentieth Century, "Ralph Bingham," accessed June 5, 2022, https://digital.lib.uiowa.edu/islandora/search/Ralph%20Bingham?type=edismax&cp=ui%3Atc.

5. A graphic representation of this stereotype, thematically close to "Elder Eatmore's Sermon on Throwing Stones," is the Currier and Ives print "A Surprise Party" (publ. ca. 1883), by the popular artist Thomas Worth. This print is part of the heavily racist Currier and Ives Darktown series. It depicts a Black parson, concealing a freshly stolen turkey behind his back, confronting two of his parishioners emerging from a poultry coop with turkeys ineffectively concealed under their coats. The caption reads, "Ah! habn't I done tole yer not to covert yer neighbor's fowl?" See Gale Research Company (1983), no. 6368.

6. There is an extensive literature on nostalgia for the old South in which the plantation is portrayed as idyllic, slavery as benevolent, and Black people as simple, docile, acquiescent, and potentially entertaining. For a review of the literature and the field, see Anderson 2005.

THREE

"WE ALWAYS ENJOY A GOOD STORY"

From Monologue to Audio Theater

INTRODUCTION

Thomas Edison, we know, enjoyed a good story. While storytelling did not figure in his—or anyone else's—public conjectures concerning the ways of speaking worthy of being recorded, preserved, or collected by means of his wondrous new talking machine, there were at least two documented occasions on which Edison was recorded telling what was apparently a favorite story for presentation to limited audiences. In one instance, in response to an invitation to provide a demonstration cylinder for attendees at an 1898 electrical exposition in Philadelphia, he sidestepped the request to record something pertaining to "dry scientific subjects," providing instead a recording of a tall tale, which he was convinced would be far more engaging to the audience.

Edison's telling of "The Liver Story" is a retelling of a story recorded and sent to him by a correspondent from California, who apparently deemed storytelling worthy of preserving and passing on to the great inventor himself.[1] And there are still other accounts of private individuals compiling collections of storytelling on cylinders for their own pleasure and to share with other interested parties (found in vol. 3 of *Phonoscope*, 1899). Clearly, there was a recognition among people who used the new technology before the advent of commercial recording that compiling collections of stories was a worthwhile use of the talking machine. It is not at all surprising, then, that when the record companies undertook the production and distribution of records for a commercial market, storytelling was one of the forms of talk they included in their catalogs. The question I want to address in this chapter, then, is how oral storytelling—as with sales pitches, political speeches, and sermons—was adapted to the new medium. How did the process actually work? What aspects of their history did the recorded performances carry with them as they were recontextualized from live performance to commercial recording? What new features did recording call forth? And what was the broader cultural field in which these transformations occurred? With these questions in view, I examine a series of recordings made by some of the earliest featured performers of the new medium, performers whose recordings demonstrate significant transformations that reshaped oral storytelling as it was adapted to commercial recording.

CAL STEWART, "JIM LAWSON'S HOSS TRADE"

The first example I consider, "Jim Lawson's Hoss Trade" (1903), was recorded by one of the earliest stars of the new medium, Cal Stewart. Stewart's recorded performances in his adopted persona of Uncle Josh Weathersby of Pumpkin Center, a fictional small town in rural New England, were immensely popular from the last years of the nineteenth century through his death in 1919 and for some years thereafter.[2] Uncle Josh, as Stewart portrayed him, was an incarnation of a stock figure in nineteenth-century American popular culture, the comic rustic Yankee, a characterological figure who was a symbolic vehicle for the representation of social contrasts between old-fashioned and modern social types and of the encounter of passing ways of life with the new ways of emergent modernity (Gal and Irvine 2019, 138–153; Rourke [1931] 1959, 3–32; Tandy [1925] 1964, 1–19). The stage Yankee appeared in a number of performance vehicles, including prominently dramatic sketches of rural and small-town life, rendered in a distinctive vernacular register that was characterized by stereotypical dialect forms, homely images, colorful exclamations, and an understated but loquacious mode of delivery. One of the popular guises in which the stage Yankee appeared, from the mid-1820s onward, was as a performer of stories and recitations in the form of a comic monologue (Nickels 1993, 76–77), a platform format well suited to that mode of traditional storytelling that involves virtuosic performance by a narrator who holds the floor for an extended period. While Stewart was an experienced actor in Yankee theatrical roles, the comic monologue was the performance form in which he excelled and which he cultivated with enormous success in the persona of Uncle Josh Weathersby. The overwhelming majority of Cal Stewart's recorded performances as Uncle Josh are in narrative form, but for present purposes, I focus on one of his popular recordings that illustrates one important presentational mode in the remediation of oral storytelling to phonograph records.

"Jim Lawson's Hoss Trade" is a version of a tale reported to be one of Abraham Lincoln's favorites, part of a large class of narratives in American oral tradition revolving around the horse trader as a trickster figure whose quick wit and glib tongue allow him to manipulate less clever victims to his own advantage (Bauman 1986, 11–32; Dorson 1959, 71).[3] In Cal Stewart's version, the story becomes a local character anecdote featuring Jim Lawson, one of the more colorful citizens of Pumpkin Center and the key figure in a number of Stewart's Uncle Josh routines.

Cal Stewart, "Jim Lawson's Hoss Trade" (Monarch 1475, https://purl .dlib.indiana.edu/iudl/media/534f661toq), Apr. 27, 1903

Jim Lawson's Hoss Trade, by Mr. Cal Stewart. [Laughs]
Well, sir, Jim Lawson,
    he was calculated to be just about the best hoss trader in Punkin Center.
But a gypsy come along one day
    an' I guess he took all the conceit outa Jim in the way o' tradin' hosses.   5
[Laughs]
You see, he trade 'im a mighty fine lookin' animal
    only had one very bad fault:
any time he'd go to ride 'im,
    happened to touch 'im on the sides,
        then he'd squat right down.   10

Well, Jim knowed if he didn't get rid o' that hoss,
    his reputation as a hoss trader was forever gone.
So he went over to see Deacon Witherspoon.
Deacon was old an' gouty,
    kinda hard for 'im to get around,   15
    an' he was mighty fond o' goin' a'huntin.
He had to hunt on hossback.
An' Jim says, "Deacon, I got a hoss you oughta have.
He's a setter."
Deacon says, "Why, Jim, I never heered tell o' such a thing in all o' my life. 20
Idea of a horse bein' a setter.
Bring 'im over, Jim,
I'd like to see 'im."
[Laughs]

Well, Jim took the hoss over, an' they went out a'huntin'.
They was a'ridin' along an' Jim, he saw a rabbit a'sottin' in the bushes   25
    an' he just touched the ol' hoss on the sides,
        an' he squatted right down.

Deacon says, "Well, what's the matter with your hoss, Jim?
Look what he be a'doin'."
Jim says, "Keep still, Deacon.   30
Don't you see that rabbit over thar in the bushes?
Ol' hoss is a'sottin' of 'im."
[Laughs]
Deacon says, "Well, well, I wanta know,
    most remarkable thing I ever seen in my life.
Well, now, did you ever?   35
How would you like to trade?"
'N' Jim says, "I'll trade you, Deacon,
    hundred 'n' fifty dollars to boot."

Well, they traded hosses,
    an' when they was a'comin' home,   40
they had to ford the creek back o' Punkin Center.
Well, when the ol' hosses was wadin' along through the water,
    Deacon went to pull 'is feet up,
    keep 'em from gettin' wet,
    'n' he touched the ol' hoss on the sides,   45
    and 'e squatted right down'n the creek.
[Laughs]
Deacon says, "Now, lookee here, Jim.
What's the matter with this here hoss?
He ain't a'settin' now, be he?"
Jim says, "Yes, he is, Deacon.   50
He sees fish in the water.
He's trained to set for suckers same as for rabbits, Deacon."
[Laughs]

As featured on the recording, "Jim Lawson's Hoss Trade" is told as if
to a co-present interlocutor in a face-to-face, immediate interaction. In the
opening line of the narration, Uncle Josh addresses his interlocutor as "sir,"
the conventional appellative of respect. And in line 6, the second-person
pronoun in "You see" serves as a phatic gesture that picks out the addressee
and signals that the narrative is directed at him. These deictic forms serve
as contextualization cues that invite the listener to the recording to assume
the participant role of targeted addressee, eliding the temporal gap and
technological mediation between the recording event and the listening
event.

Uncle Josh delivers the story in his characteristic register: informal, vernacular, marked by stereotypically "rural" dialect forms. Stewart, who was

born and raised in Virginia, made some effort to sound like a New Englander on his Uncle Josh recordings, but even those features that he took to be distinctive to that region extend more broadly across the country. Stewart apparently considered the preterite form "sot" for "sat" or "set" to be especially distinctive of Yankee speech, and he used it heavily, not only in this performance but in others. Indeed, he overused it, coining grammatical variants ("sottin'") that were not current in actual speech. Yet "sot," as well as "knowed" (for "knew") and "heared" (for "heard") had a wide currency in American vernacular speech (Atwood 1953, 21, 17, 16; Wentworth 1944, 580, 344, 283). Perhaps the most distinctive New England usage to mark Uncle Josh's speech is "be" as a progressive-aspect auxiliary verb for "am," "are," "is," as in "Look what he be a' doin'" (line 29), yet it too was occasionally documented in other parts of the country (Atwood 1953, 27; Wentworth 1944, 45–46).

In addition to the general dialect features that characterize Uncle Josh's speech, there are several elements that came to define his personal style, including the relatively flat intonation and the trademark laugh that punctuated the narration on all of his recordings. In all, however, while Uncle Josh's speech displayed a few features that would have been recognizable indicators of New England dialect to those familiar with the region, his overall speech style was an amalgam of lexical and grammatical elements of relatively broad regional distribution in American rural vernacular speech that listeners throughout the United States would have recognized as a generalized "country" dialect, perhaps not precisely like the dialect of their own region but with enough familiar elements to do the indexical work of establishing Uncle Josh as a quintessentially rustic figure.

The formal structure of the narrative exhibits many of the common features of American oral storytelling, including prominently the following:

1. The use of "well" as an opening and subsequently as an episode marker (lines 11, 24, 39, 42) (cf. Bauman 1986, 23, 25, 42, 56–58, 83–90, 108–111)
2. The division of the narrative into three principal episodes consisting of dyadic interactions between the dramatis personae (here Jim Lawson and Deacon Witherspoon), played out in quoted dialogue (cf. Johnstone 1996, 40)
3. The predominant use of the historical present ("Jim says," "Deacon says") of the verbs of saying employed as quotative frames (cf. Johnstone 1990, 77–88)
4. Uncle Josh's signature laugh, an exuberant cackle that signals the narrator's own amusement at the narrated event and serves

as an evaluative marker of the key element on which the plot turns—namely, the squatting of the horse

5. The clever, reportable final utterance, which has the effect of a capping punch line, a characteristic formal feature of the oral anecdote (cf. Bauman 1986, 54–77)

6. The polysyndetic, additive chaining of narrative clauses, present here (lines 25–27) but more common in other stories (cf. Johnstone 1990, 50–58; Schiffrin 1981, 131–132)

All of these features—the deictic contextualization cues, the vernacular speech styles, the formal organization of the narratives—impart some of the immediacy of face-to-face vernacular storytelling to the recorded performance of "Jim Lawson's Hoss Trade" as it is played back in the listening event. To be sure, the recorded performance is conspicuously fluent and devoid of any signs of audience participation, but the most virtuosic of traditional oral storytellers do approach or even attain this level of fluency, and the organization of participation in such performances does allow the performer to monopolize the floor while limiting the audience essentially to back-channel responses. The listener can still respond at a distance with a smile or a laugh, and Uncle Josh's own trademark laugh provides a functional substitute for other participants' responses.

An additional mediating element is the introductory announcement that opens the recording, transcribed in line 1: "'Jim Lawson's Hoss Trade,' by Mr. Cal Stewart. [Laughs]." Announcements of this kind were common on early recordings and served, among other functions, as orienting frames for the recorded performances (Feaster 2006, 304–346). Here, by giving the recitation an objectifying title, by identifying himself by his own name (not Uncle Josh), and by including his trademark laugh, Cal Stewart frames the narration to follow as a simulation, a reenactment, of a traditional, oral storytelling performance. This framing is entirely consistent with Stewart's public characterization as an "impersonator" or "delineator" of the comic Yankee type. The announcement, and Stewart's identification as the animator of Uncle Josh, serve a mediating function, framing the storytelling performance as a constructed display object, albeit one that is a skillful and persuasive simulation of the real thing.

To be sure, however, the platform performances of the Yankee storyteller on which Cal Stewart modeled his Uncle Josh recordings were already simulations, representations of traditional storytelling.[4] The production framework of oral storytelling performance—extended holding of the floor, phatic gestures to a co-present gathered audience—lends itself readily to platform performance and allows quite well for the retention of the core features of traditional storytelling we have identified. But the crafted rustic persona, with his stylized, stereotyped dialect, are artifacts of the stage Yankee, a symbolic construction of nineteenth-century American popular culture. "Jim Lawson's Hoss Trade," then, is a mediated representation of a stage representation of oral storytelling.

EDWIN WHITNEY, "THE FARMER AND THE HOGS"

If we consider "Jim Lawson's Hoss Trade" to represent a second-order transposition of oral narrative performance, insofar as it involves the adaptation to sound recording of a platform format that is itself a representation of a traditional face-to-face storytelling performance, our second example, "The Farmer and the Hogs," adds still other layers of mediation to the performance. This recording, a rendering of a traditional American tall tale,[5] begins with a bit of metanarrational contextualization by the performer, Edwin Whitney, reporting and evaluating a "mighty good" story that is part of the active repertoire of another raconteur, one Strickland W. Gillilan, identified as a "funnyman."[6] Whitney thus offers a number of warrants for telling the story again: it sustains multiple tellings by Gillilan, Gillilan is known for being funny, and it is "a mighty good story" to boot. In functional terms, these devices both authorize and traditionalize the story, aligning Whitney's performance to follow with antecedent performances.

Edwin Whitney, "The Farmer and the Hogs" (Victor 16489B, https://
purl.dlib.indiana.edu/iudl/media/920f85ts7z); recorded June 15, 1909

A mighty good story is told by Strickland W. Gillilan,
   the *Baltimore American* funnyman,
better known to you possibly as the author of
"Off Agin, On Agin, Gone Agin Finnigan."
He tells the story of a gentleman who was traveling in the state of Arkansas. 5
He was on horseback,
   going from one village to another,
and while riding past a piece of woods,
he noticed a number of hogs,
all acting very strangely.   10
He stopped and watched 'em.
The hogs would all stand still,
   tuck up their ears and listen,
and at the same time would all start and run in the same direction
   as fast as they could go.   15
They would run a little ways,
   stop,

listen,
run again in a different direction.
Run a little ways,   20
stop,
listen,
and run some more,
and they kept at it.

The gentleman's curiosity was so much aroused   25
that he stopped at the first cabin he saw
   as he rode on
   and greeted the old man who came to the door.

The old man replied,
[hoarse, gravelly voice; slowly] "Howdy, stranger."   30

"Now, my friend,
   have, uh, you lived here all your life?"

"No, sir."

"Well, maybe you've lived here long enough to be able to tell me
   something about who owns this piece of woods up the road,   35
   the first piece on my right."

"Yes, sir,
I reckon I can tell you."

"Well, who owns it, please?"

"I do."   40

"Well, those must be your hogs in the woods, then,
   are they?"

"Yes, sir,
   I reckon those're my hogs."

"Well, I wish you could tell me what's the matter with them.   45
I've been up there watching 'em for the last half hour,
   and I never saw anything act the way they do.
They all stand and listen,
then start 'n' run
   all in the same direction at the same time.   50
They run a little ways,
   stop,
   'n' listen,

run again in the opposite direction perhaps.
Run a little ways,   55
    stop,
    'n' listen,
    'n' run again,
    stop,
    listen,   60
    and run some more,
always in a different direction,
and they keep at it.
Can you explain to me
    why they do that?"

"Yes, sir,   65
I reckon I kin tell you.
You see, stranger,
this is the season o' the year
when we're fattenin' up the hogs
    and gittin' 'em ready for market.   70
Well, I been a' goin' out 'n'
    callin' 'em up to the fence,
    throwin' over a little corn to 'em
    as fur as it gets 'em
    started along.   75
Here about two weeks ago,
I got this cold,
and it settled right on my vocal cords
so I cain't talk.
I can't holler   80
    and call the hogs no more,
so I had to go out 'n'
    sound on a tree with a stick to call 'em up.
Well, they got so they'd answer that all right.
An' now these dumb woodpeckers,   85
    they keep 'em crazy.

Whitney's telling is framed initially as a report of Gillilan's tellings, but Gillilan soon recedes from the scene as Whitney takes over the responsibility for the current performance. This is a common move in traditional story-telling: establishing one's authority for the current performance and claims on the audience's attention by intertextual linkage to antecedent authoritative performances and then taking over the performance oneself. I have more to say about the attribution to Gillilan a bit later. For now, however, I would note that like Cal Stewart's "Well, sir" at the opening of "Jim Lawson's Hoss Trade," Whitney's suggestion that Gillilan is "better known to *you*" as the author of a widely circulated poem can plausibly be heard, I think, as an invitation to the hearer of the recording to assume the participant role of targeted addressee.

The narration itself begins by setting up the symbolic tension that will provide the central dynamic of the story—namely, the classic confrontation between the cultivated (often urban, often eastern) outsider and the frontier rustic (Dorson 1959, 92). One inflection of the dynamic that drives these symbolic encounters is a turning of the social tables, as the sophisticated outsider is baffled by a feature of the rustic environment that is completely beyond his ken but readily explainable by the backwoods farmer, even though the explanation may turn out to be a tall tale. Here, we have a "gentleman" traveling through Arkansas, the quintessential region of backwardness and rusticity, and encountering a herd of pigs behaving in a strange and puzzling manner, the complicating action of the story. All this narrative setup is accomplished by past-tense description of the narrated event, which continues to the point when the gentleman stops at the cabin and greets the old man who came to the door.

Then, with the old man's return greeting, framed by the quotative frame "The old man replied," the narrative framework shifts to direct discourse, from diegetic to mimetic modes of performance. The entire remainder of the story is rendered as quoted speech, with the attendant deictic shifts in pronouns, tense, demonstratives, and the like. What is especially striking about the dialogic interaction in this performance, however, in marked contrast to "Jim Lawson's Hoss Trade," is that it is accomplished without quotative attribution, but it remains abundantly clear nevertheless who is speaking throughout the remainder of the tale. The overall effect is to foreground the mimetic mode of presentation, drawing the performance closer to theatrical enactment than to narrative exposition. It is crucial to recognize, however, that unlike staged theatrical performance, which combines verbal and visual semiosis, the recorded performance is restricted entirely to the acoustic channel. When dialogic interaction is enacted onstage, the audience can see as well as hear who is speaking at any given time. In the recording of "The Farmer and the Hogs," Whitney must rely on speech alone. He manages the narrative dialogue by a number of expressive means that build up and sustain the contrast between the dramatis personae.

1. The two interlocutors are differentiated, first of all, by contrastive speech registers: the gentleman speaks standard English, marked by formal syntactic constructions, politeness forms,

phonologically standard pronunciation, and carefully marked
word boundaries, while the farmer speaks a rural dialect, distinguished by stereotypically rustic grammatical constructions
and phonology and by honorifics that acknowledge his interlocutor's gentlemanly status.

2. The two speakers are further distinguished by voice quality and
   expressive style. The gentleman's voice is well modulated, while
   the farmer speaks with a husky, raspy timbre, a consequence, we
   discover, of a cold that "settled right on my vocal cords so I cain't
   talk." In a related vein, the farmer tends toward laconicity of expression, at least in his early turns at talk, as he responds to the
   relatively voluble solicitations of the gentleman "stranger."

3. The contrastive ways of speaking that distinguish the two interlocutors are further augmented by the structure of the conversation they play out, their respective turns both differentiated and
   tied together by a range of cohesion devices:

   • question-answer adjacency pairs: "Well, who owns it please?"
     "I do."
   • yes or no responses that are tied to antecedent utterances by
     assuming a positive or negative alignment to them: "Now, my
     friend, have you lived here all your life?" "No, sir."
   • lexical repetition: "Well, those must be your *hogs* in the
     woods, then, are they?" "Yes, sir, I reckon those're my *hogs*."
   • syntactic parallelism, as the old man responds with variants
     of the gentleman's solicitational questions: "Well, I wish you
     could tell me," "Yes, sir, I reckon I kin tell you."

All these devices are found in traditional oral storytelling, but I cannot recall
a performance, even by the most virtuosic of storytellers, in which quotative frames are so systematically and completely dispensed with and direct
discourse managed with such a complex inventory of devices to differentiate the voices of the dramatis personae. As I suggested a bit earlier, part of
this strong effort at differentiation is attributable to the exclusive reliance in
recordings on sound alone, which must bear all the burden of keeping the
characters clear.

Medium alone, however, cannot account fully for the performance style of
"The Farmer and the Hogs." To bring the style of the recording more clearly
into focus, it is illuminating to consider more specifically what performance
forms underlie Whitney's recorded representation. This requires first a word

about tent Chautauqua. The Chautauqua movement developed in the period following the Civil War originally to serve the needs of Methodist Sunday school teachers and other active church members by providing a summertime venue for teacher training, Bible study, spiritually edifying lectures, devotional exercises, and other vehicles of spiritual cultivation. The institution expanded through the late nineteenth and early twentieth century in the form of touring tent shows coordinated by regional or national booking agencies working in tandem with local organizers, with offerings that included—beyond the original spiritually oriented forms—musical and theatrical performances, humorous lectures, and variety entertainments (Canning 2005; Tapia 1997). During this period, Chautauqua converged in various ways with the older institution of the lyceum, founded in the antebellum period as a movement to improve the intellectual and moral culture of small towns and rural communities, originally in New England but soon extending to the rest of the country (Ray 2005). Lyceum programs featured lectures, debates, classes, and other edifying presentations in public spaces, including purpose-built lyceum halls. After the Civil War, the lyceum circuit, like Chautauqua, extended its scope to include various forms of popular entertainment, and both institutions relied on the same booking agents for performers. Eventually, the two convergent institutions developed a seasonal complementarity, with Chautauqua tent shows touring in the summer and lyceum programs offered in lyceum halls during the colder months.

Although Whitney identifies Strickland W. Gillilan, the source of his story, as a journalist and author of popular poetry, Gillilan was also an active performer on the Chautauqua and after-dinner circuits, delivering humorous "entertainment lectures" on a variety of subjects. One of the lectures he offered to prospective clients was "A Sample Case of Humor," in which he essayed a definition of humor and a typology of humorous themes, "illustrated with stories." Among his stock themes was "Humor of Rusticity," for which "The Farmer and the Hogs" might well have been one of the illustrative examples. Gillilan, then, like Cal Stewart, employed the format of a platform performer, but whereas Stewart performed in the guise of a stage Yankee, animating a rustic storyteller, Gillilan delivered lectures, using stories for illustrative purposes. Tent Chautauqua was conceived and framed as an educational institution, a means of bringing culture and learning to the rural parts of the country (Canning 2005; Tapia 1997). The Chautauqua lecture was formally marked by smooth and showy fluency, without hesitation phenomena, false starts, or repairs, and by a tendency—passages of quoted speech aside—toward phonologically and grammatically careful standard English.

Like Gillilan, Edwin Whitney was an active performer on the Chautauqua circuit, known for, among other things, his dapper, fastidious dress (H. Harrison 1958, 103–105; Tapia 1997, 138–139). More importantly, though, for our purposes, Whitney was a skilled "monactor" or "monodramatist," a specialist in a distinctive mode of Chautauqua dramatic performance in which an individual animated all the characters in a play or dramatic sketch (Case and Case [1948] 1970, 54–55; Rieser 2003, 231–232). Whitney was advertised in a program brochure as being able to voice "a whole play company at once." It is these Chautauqua skills that we hear on "The Farmer and the Hogs." The opening metacommentary, with its intertextual allusion to Gillilan's story, is in lecture register, which extends into the early diegetic exposition: smooth, fluent, Standard English, though with a few vernacular touches. The shift into direct discourse marks the transition from diegetic reporting to mimetic enactment, and this latter portion of the performance is in the monactor mode of animating all the dramatis personae.

The monactor mode of dramatic performance was largely a concession on the part of Chautauqua promoters to the religious sensibilities of their core constituencies, who had deep moral reservations against full, staged theater. Having one person—not in costume, not a morally suspect actor—voice all the parts of a classic or properly uplifting play could be promoted as modeling skilled elocution and thus warranted as consistent with the intellectual and moral goals of Chautauqua. Adapted to storytelling, the same mode of dramatic performance became a virtuosic, formalized, theatricalized adaptation of the taking on of voices in direct discourse that is a common feature of traditional vernacular storytelling. Moreover, it is readily adaptable, without significant modification, to sound recording. Edwin Whitney's performance of "The Farmer and the Hogs," like Cal Stewart's "Jim Lawson's Hoss Trade," is thus a remediated representation of a mode of storytelling characteristic of tent Chautauqua, adapted to more extended lecture presentations and marked by theatricalized voicing of narrative dialogue in a manner drawn from one-person dramatic readings or monactor performances.

LEN SPENCER, "UNCLE JIM'S RACETRACK STORY"

My final example, "Uncle Jim's Racetrack Story," has presentational affinities with "The Farmer and the Hogs" in certain respects but represents an entirely new departure in the remediation of storytelling. The recording was made by Len Spencer, one of the first, most versatile, and most successful of the early commercial recording performers (Gracyk 2000, 314–319). Spencer's father operated a business college in Washington, DC, where the son worked as an instructor. The college made use of recording machines as instructional tools for training in public speaking, and Spencer had frequent occasion to visit the nearby headquarters of the Columbia Phonograph Company to have the machines serviced and to replenish the school's supply of cylinders. On one of his visits, Spencer expressed his desire to have his own voice recorded and turned out to have a strong baritone that was well suited to the recording technology of the day. He began his recording career around 1889 or 1890, making song recordings for commercial coin-operated machines that were placed in train and ferry terminals, arcades, and other public places. That is to say, Len Spencer was from the beginning a phonograph performer. Unlike Cal Stewart or Edwin Whitney, his entire career was linked to the new technology of sound recording. Thus, while he was clearly familiar with the popular entertainment forms of his day, including the monactor mode of performance in which one performer animates all the characters in a dramatic enactment, he was more professionally attuned to the performance potential of the new communicative technology.

At the heart of "Uncle Jim's Racetrack Story" is an enacted representation of a storytelling performance situation in which the narrator relates a story to an audience whose members contribute back-channel responses and evaluative metacommentary. The story itself is an exemplar of another genre like the tall tale, which represents a playful, rekeying transformation of a personal-experience narrative. Uncle Jim's story is a catch tale, which derives its effect by drawing the audience to a particular interpretive expectation, only to pull the rug out from under them at the end by deflating and discrediting their assumptions. As we will see, audience response is essential to the narrative exchange represented on this recording: without the expression of mistaken understanding, Uncle Jim cannot produce his rekeying response.

Len Spencer, "Uncle Jim's Racetrack Story" (Victor 2790, https://purl .dlib.indiana.edu/iudl/media/g64t54s418), Apr. 21, 1904[7]

[Hoofbeats]
Jim: Whoa! Whoa! Whoa!
Get over there!

Host: Hello, Jim.
Why, uh, here's some friends of mine from New York.   5
I'm just, uh, showing them through the paddock
and I want, uh, you to tell them that story about the Tennessee Derby.

Guest: Go ahead, Jim,
we always enjoy a good horse story.

[Bugle call]   10

Jim: You durn fool!
Whoa, there! Whoa! Whoa!
Durn that horse.
Ol' Marster Dan'l entered Possum in the Tennessee Derby a down in
    Memphis.
I was only a stable boy, them days.   15
[Band starts to play]

Guest: Yessir.

Jim: Just before the race,
while the band was a'playin'
just like you hear it out yonder in the grandstand now,   20
the old master run in and he says, "Jim,
Little Pete just done broke his leg in the last race,
And, uh, you'll have to ride Possum.
Now don't stand there lookin', boy,
but get into them colors quick,   25
and here's, uh, your instructions:
Now, uh, there's only one horse I'm afraid of in the derby;
that's Arizona.
Watch 'im!
Lay close to 'im, boy, till the stretch,   30
and them climb up and beat 'im up!
If you don't, I'm a ruined man."
Yessir.
[Hoofbeats]
Whoa! Whoa there! Whoa!   35
Durn that horse!
I never rid a race in my life,
but I knowed how to watch the flag,
and this time up
I seen the yellow flag swish down   40
and we was off!
[Crowd cheers; rapid, multiple hoofbeats]
You couldn't hear nobody's step [?]
for yellin' in the grandstand
just like they are there now.   45

At the first pole, Arizona was a length ahead of Possum.
[Crowd cheers; music stops]

Guest: Yes, yes, well?

Jim: Possum was a'goin' easy, pullin' for 'is head.
At the second pole, Possum was right behind Arizona,   50
still a'pullin' for 'is head.

Guest: Yes, yes, well?
[Crowd cheers]

Jim: In the back stretch, Ol' Buck come up on the outside,
and then Possum was in the pocket,   55
with Arizona five lengths ahead.
[Crowd cheers]

Guest: Gee, whiz!

Jim: So we turned to the three quarters.

Guest: Yes?   60

Jim: Then we piled into the stretch.
I worked Possum up on the outside and let 'im go.
[Crowd cheers]

Guest: Good ploy, Jim!

Jim: Tear up toward the stands.   65
[Crowd cheers]

Guest: OK.

Jim: All together like mad.
[Crowd cheers]

Guest; Yes? Yes?   70

Jim: Arizona two lengths ahead.

Guest: Well, go on!

Jim: I give Possum the whip.
Go there!

And I'm in!   75
[Crowd cheers]
Four lengths ahead of Arizona.
[Crowd cheers]

Fig. 3.1. Horse race featuring an African American jockey, 1892. Library of Congress Prints and Photographs Division, Washington, DC.

Guest: Then you won the race!

Jim: Heh. Aw, no, sir.   80
There was—heh!—four-five other horses ahead o' Arizona.
[Laughter; bugle call; hoofbeats stop]

While "Uncle Jim's Racetrack Story," like "The Farmer and the Hogs," uses the monactor presentational mode, this performance exploits a more extensive range of semiotic means and formal devices for dramatic ends and achieves, by these means, a far more complex sonic representation than any of the other recordings we have examined. Consider, to begin with, what a wealth of theatrical business is accomplished in the first twenty lines of the transcript, solely by acoustic means. Let us start with the mise-en-scène. The opening sound effect of hoofbeats, the command "Whoa! Whoa! Whoa! Get over there!," the command plus exclamation "Whoa, there! Whoa! Whoa! Durn that horse," and the explicit reference to "the paddock" evoke a man on horseback in the paddock area of a racetrack. Voice changes, marked by a dialect shift and pronominal deixis; the greeting, "Hello, Jim"; and the reference to "friends of mine from New York" round out the dramatis personae: the rider is Jim, an African American jockey, and his interlocutors are a White racing man hosting two or more White friends from New York. The band music that begins to play around line 16 serves as a further sound effect that allows for additional spatial contextualization: the low volume of the music and the adverbial phrase "out yonder in the grandstand" expand the mise-en-scène to include additional and more spatially distant features of the racetrack complex. These devices set the stage for the storytelling performance, which is elicited by the host in line 7: "I want you to tell them that story about the Tennessee Derby," marking the narrative as a part of Jim's repertoire, a set piece that he had told on other occasions.

The narration itself begins on line 11, as Jim offers some characteristic orientational information concerning the narrative event to follow, an account of his first race as a jockey. Drafted at the last minute to substitute for a rider who has broken his leg in the preceding race, Jim takes on the voice of his master in direct discourse, as his master coaches him in the strategy for his horse, Possum, to beat the rival, Arizona, in the Tennessee Derby. In terms of the recorded performance, this bit of direct discourse represents a double lamination, with Spencer voicing Jim voicing his master. Jim's affirmative response, "Yessir," marks the end of this first episode, followed by a frame break, as Jim steadies the horse on which he is mounted in the ongoing storytelling event. Returning to his account of the Tennessee Derby, Jim offers a bit more orientational information concerning his readiness for the race and then launches into an increasingly suspenseful account of the race itself, from start to finish. The tension is provided by the shifting dynamics of the race and his closing drive to the finish line, four lengths ahead of the rival horse, Arizona.

Jim's narrative performance is punctuated by several effective metanarrational devices. No sooner does the starter's flag drop and the race begin in his account, when other contextualizing sound effects of the mise-en-scène link the narrative event to the narrated event, intensifying both. The crowd in the grandstand, in the background of the narrative event, cheers in the distance (indicated by the low volume of the noise), and Jim draws that noise into his account by linking it to the corresponding cheers of the crowd at the Tennessee Derby when his first big race began (lines 41–44). At the same time, multiple hoofbeats sound in the distance as a race in the background of the narrative event proceeds. As the Tennessee Derby progresses and the narrative suspense intensifies, the back-channel responses of Jim's audience grow more and more excited and come at shorter and shorter intervals, urging him on to the climax as Possum gains on Arizona. The effect is further intensified by the cheering of the crowd in the background of the narrative event, which serves at the same time to punctuate Jim's story. Then the climax: "And I'm in! Four lengths ahead of Arizona." "Then you won the race!" exclaims an excited guest. But no—here's the catch: "There was four-five other horses ahead o' Arizona."

"Uncle Jim's Racetrack Story" thus emerges as a small piece of audio theater (Feaster 2007, 375–389), a dramatic representation of a narrative event that contains a storytelling performance. All the dramatic effects of this little enactment are acoustically crafted; that is to say, the recorded performance is more than a simple matter of placing a performer in front of a recording apparatus. Rather, the performance is the product of sound design, an essential quality of audio theater. Sound design is the crafting of acoustic resources in an audio production to create or evoke features of characterization, aspects of social relations and interactional alignment, configurations of space and location (including point of view—or better, of audition), vectors of motion, and the dynamics of affect and atmosphere. Let's examine a bit more closely the elements and organization of the sound design that shapes "Uncle Jim's Racetrack Story."

Clearly, the primary semiotic resource exploited in the sound design of the recording is speech, which serves multiple functions in the crafting of the performance. It contributes, as I have noted earlier, to the setting of the scene: recall, for example, the way that Jim's commands allow us imaginatively to picture him on horseback. Or consider how the host's references to the paddock or Jim's to the crowd "out yonder in the grandstand" establish the proximal and distal features of the racetrack complex in which the narrative event takes place.

In addition to the evocation of setting, speech serves in a variety of ways as an instrument of characterization. Absent any visual means of establishing the cast of characters as embodied individuals present before us on a stage, speech has to carry the entire burden: by address ("Hello, Jim"), reference ("some friends of mine from New York"), aural contrast (African American vernacular English versus standard English as a means of distinguishing Jim from the host and his guests, or lower- versus higher-pitched voices as a means of distinguishing the host from a guest), or other means. Likewise, speech alone must serve as the means of establishing and inhabiting participant roles, without gesture, proxemics, gaze, or other visible signs to help us sort out the storytelling performer from his audience. And finally, of course, speech is the means by which Jim performs his story in the narrative event that is the centerpiece of the recorded enactment.

A critically important complement to speech in the sound design that makes up "Uncle Jim's Racetrack Story" as a dramatic enactment is the repertoire of sound effects that figure prominently in the recording. Indeed, sound effects bracket the entire recorded enactment. The first thing we hear, on playing the record, is a clopping sound that is a sonic icon of hoofbeats, which suggest in turn the presence of a horse. That indexical inference is reinforced a moment later by the commands, "Whoa! Whoa! Whoa! Get over there!" The bugle call, just before Jim begins his story (line 10), the band music that strikes up as he opens his narration (line 16), and the cheers that ring out in the background (indicated by relative volume) as he tells his story of the Tennessee Derby all serve as environmental and spatial indicators, filling out our imaginative understanding of the racing complex, its configuration, and the deictic center and periphery of the focal event within it. The sound effects are supplemented by Jim's explicit references to the band "out yonder in the grandstand" or the crowd "there now." Perhaps the most ingenious aspect of the use of sound effects in the acoustic design of the skit is the way in which the sounds of the *narrative* event—the band music, the cheers, the multiple hoofbeats in the background—merge with the sonic features of the *narrated* event, Jim's first race, to enhance the semiotic texture of his suspenseful account. The crowd cheering the riders in the background of Jim's storytelling performance seems also to be cheering him in the race he is recounting, helping to build the excitement as Possum gains length by length on Arizona.

Also impressive is the use of sound effects to punctuate the structure of the dramatic enactment and the story performance within it. The first bugle call and the second set of hoofbeats as Jim steadies his horse once again (lines 10–12) mark the transition from the dramatic setting of the scene to the commencement of Jim's narrative performance. In a similar vein, the point at which the band begins to play marks the transition from the orientation section of Jim's story to the first narrative episode, in which Jim's master drafts him to ride in the Tennessee Derby and schools him in strategy for the big race. The next set of hoofbeats and Jim's third effort to control his horse (lines 34–36) signal the shift to the principal narrated episode—namely, the Tennessee Derby itself—and the cheers of the crowd continue to mark the stages of the race as Jim ticks off his stage-by-stage gains against Arizona and, ultimately, his climactic finish as Possum comes in four lengths ahead of his rival. The final bugle call, following the responsive laughter of Jim's audience, serves as an acoustic closing curtain, framing the end of the dramatic enactment and of the recording. Sound, then, is an integral structural element of the narrative event, the narrated event, and the narrative itself in the construction of this pioneering piece of audio theater.

CONCLUSION

The three recordings I have examined in this chapter are mediated representations of oral storytelling performances. As we have seen, however, they are far from direct, transparent transpositions of traditional storytelling by vernacular storytellers in conventional contexts. Rather, our examination has revealed them to be remediations of what were already transpositions of traditional oral storytelling from sociable encounters to the produced and commodified platform events of stage monologues and Chautauqua lectures. Moreover, the recordings were made by professional performers, in carefully fitted-out recording studios. And, notwithstanding the association of storytelling with a rustic, agrarian milieux that figures in the representations we have examined—the small-town cracker-barrel philosopher, the backwoods farm in Arkansas, the southern racetrack—the recordings were produced in the urban centers of New York and Washington, DC. Still, in the recontextualized and remediated form in which we encounter them on the recordings, the storytelling performances carry at least some of their contextual history with them—that is, elements of both traditional oral storytelling and the platform events from which they were transposed and adapted for the recordings.

At the center of all three recordings is the representation of the familiar act of storytelling, the verbal production of a narrative text. "Jim Lawson's Hoss Trade" and "The Farmer and the Hogs" are documented in oral tradition and display thematic, formal, and metapragmatic features that establish them clearly as exemplars of two of the most popular narrative genres in the American repertoire of oral narrative. "Uncle Jim's Racetrack Story," while not in itself traditional, is cast in a traditional genre—the catch tale, a double-voiced genre, which, like the tall tale, builds on and dislocates the generic expectations surrounding the narrative of personal experience.

Linked closely to the act of storytelling, of course, is the role of narrator and its complement of related participant roles. Each of the examples configures that participant structure in a different way. Cal Stewart animates the persona of the traditional country storyteller: Uncle Josh is a figuration of the classic rural cracker-barrel raconteur, mediated through the long-established stage figure of the comic Yankee. By the use of various devices—appellatives, pronouns, an informal register—Uncle Josh addresses his narration as if to a co-present interlocutor in a sociable encounter, inviting the listener to assume that participant role. Edwin Whitney likewise appears to be addressing a co-present audience, but whereas Uncle Josh's "sir" picks out an individual addressee, Whitney's "you" is more ambiguous. With no visual cues to setting and participation frameworks, it may be taken as second-person singular, inviting the listener to the recording to imagine him- or herself in a dyadic encounter with the narrator. On the other hand, the register and performance style of Whitney's performance evoke the more distanced participant structure of a lecture, still part of the interaction order, in which his "you" of address suggests the second-person plural of the audience as a collective. In this alignment, the listener to Whitney's recording is interpellated as a member of the gathered audience of the Chautauqua. In "Uncle Jim's Racetrack Story," Len Spencer animates Uncle Jim as another stock figure like the comic Yankee: the amiable, avuncular African American storyteller of which Uncle Remus, of course, is the most familiar exemplar. The participant structure of this recording, though, is a bit more complicated than the other two. The framing of "Uncle Jim's Racetrack Story" is theatrical: the listener to the recording is a bystander/overhearer of a dramatic enactment, and there is no direct prior experience invoked by the performance; it is like a staged theatrical drama but yet not like it. It has a plot, with multiple characters, but it is restricted solely to the auditory channel and relies on sound design for critical dramatic effects: mise-en-scène, spatialization, and other contextualizing functions. This performance is most completely the product of the new medium. The recording is a true media text: a commodified utterance addressed to no one and to everyone (Warner 2002). To be sure, the other recordings are media texts as well in this sense. My point is that they don't show it as fully and clearly.

What is at play, then, in these recordings, is the construction and manipulation of a tension in modes of address to an audience, which interpellates the listener in two different orders of public. The framing of "Jim Lawson's Hoss Trade" and "The Farmer and the Hogs" evokes the public of a co-present gathered audience. In the former, it is primarily the intimate audience of sociable storytelling, though with elements of the more distanced participant structure of platform storytelling; in the latter, platform storytelling comes more strongly to the fore. In "Uncle Jim's Racetrack Story," the listener is more clearly positioned as the overhearer/consumer of a media text: anonymous, distanced, dispersed—that is, as a member of a distributive public, constituted by the circulation of the recording as commodity. And, indeed, so are the listeners to the other two recordings, in a participant framework that is laminated onto the ones evoked by the recorded performances themselves.

In addition to the relationship between gathered and distributive publics called into play by these recordings, I would suggest that they also align the listener to at least one additional public. I have already suggested the dynamic tension implicated in the representation of oral storytelling, that most traditional of communicative practices, on phonograph records, that most modern of communicative technologies—at least in the turn-of-the-twentieth-century US. Recall that storytelling on the recordings we have examined also indexes the rural, small-town past as well as the encounter between rural folk and urban sophisticates: the gentleman traveler in Arkansas, the visitors from New York at the southern racetrack. The recordings, in essence, render oral traditional storytelling as a symbolic vehicle of nostalgia, a trope for the sentimental evocation of a folkloric past still familiar to the marketing targets of the phonograph industry: urban dwellers, bourgeoisified small-town people, and prosperous farmers for whom mechanized agriculture provided sufficient leisure and income to indulge in commoditized forms of home entertainment. Again, this mode of symbolic construction becomes explicit in the metadiscourse of advertising. Taking Cal Stewart as an example once again, as he was the most famous "talking machine storyteller" of them all, the key throughout his career was nostalgic retrospection: notices and advertisements in newspapers and trade journals characterized Stewart as "'right down tew home' among the folks," his manner as "quaint," Pumpkin Center as "romantic," and the scenes depicted in the recorded stories as "old fashioned."[8] This form of mediatized nostalgia, then, invites consumers/listeners to align themselves to a shared American past, what we might term a historically founded public. But from the first invention of the idea of folklore, in the eighteenth century, it has always served well as a symbolic vehicle for constructing links to a shared, primordial past, as part of a modernizing and nationalizing vision. The remediation of storytelling on early commercial recordings, then, enlisted a new communicative technology in the service of a time-honored mechanism of modernization, the re-creation of a nostalgic vision of the past. The recorded performances I have discussed in this chapter, that is to say, are the popular, performative, commoditized analogs of the more philosophical efforts of social theorists to folklorize vernacular forms of expression. They are perfect instruments of bourgeois nostalgia—buy a simulacrum of the past that will nevertheless allow you to display how very modern you are.

NOTES

1. The manager of a recent electrical exposition in Philadelphia asked Edison to send on a phonographic cylinder setting forth some of his latest ideas of electrical interest. Edison complied in his own way. The message was as follows:

My Dear Marks: You asked me to send you a phonographic cylinder for your lecture this evening and to say a few words to the audience. I do not think the audience would take any direct interest in dry scientific subjects, but perhaps they might be interested in a little story that a man sent me on a phonographic cylinder the other day from San Francisco. In the year 1873, a man from Massachusetts came to California with a chronic liver complaint. He searched all over the coast for a mineral spring to cure the disease, and finally he found, down in the San Joaquin Valley, a spring, the waters of which almost instantly cured him. He thereupon started a sanitarium and people all over the world came and were quickly cured. Last year this man died, and so powerful has been the action of the waters that they had to take his liver out and kill it with a club.

Yours truly

EDISON ("Our Tattler" 1898b, 13)

2. For biographical information on Cal Stewart, see Feaster 2006; Gracyk 2000, 332–338; McNutt 1981; Petty 1974; Walsh 1951a, 1951b, 1951c, 1951d.

3. Motifs K134 Deceptive horse sale or trade (Baughman 1966, 341), K134.6 Selling or trading a balky horse (Baughman 1966, 342).

4. On the notion of platform performance, see Goffman 1981, 165–166; and Wilson 2006, 59–94.

5. Motif X1206(ba) Owner, unable to call hogs because of sore throat, calls them by tapping on fence board with stick. The hogs chase through the woods to various spots where / woodpeckers are drilling, hoping for more food (Baughman 1966, 476).

6. For biographical information on Strickland W. Gillilan, see H. Harrison 1958, 174–179; and Tapia 1997, 68–69. An excellent source of materials on Gillilan's career as a Chautauqua performer, see the University of Iowa library's web archive, Traveling Culture: Circuit Chautauqua in the Twentieth Century, "Spencer Gillilan," accessed June 5, 2022, https://digital.lib.uiowa.edu/islandora/search/spencer%20gillilan?type=edismax.

7. In the transcriptions in this chapter, I have added two additional graphological devices to those outlined in my prefatory note on transcriptions to highlight still further the poetic patterning principles that organize the storytelling performances. As elsewhere, line breaks mark breath units, intonational units, or syntactic structures, which are usually—though not always—mutually aligned. In addition, indented lines mark shorter pauses, and double spaces mark episode breaks or changes of represented speaker in direct discourse.

8. These quotes are drawn from the following sources: EPM, September 1908, 26; *Coshocton Daily Age*, March 4, 1902, 1; EPM, September 1908, 26; EPM, January 1909, 18.

FOUR

—ᴍ—

"TALKING MACHINE STORYTELLER"

Cal Stewart and the Remediation of Storytelling

IN THE EARLIER CHAPTERS OF this book, my primary focus has been on how a range of performance forms and practices that participants were accustomed to engaging in situations of co-presence were adapted to the new communicative technology of sound recording during the formative period of commercial record production in the US. I have examined how sales pitches, political speeches, sermons, and stories, each with its own distinctive generic qualities, were transformed in the process of remediation, a process that involved the recontextualization of these performance genres and their associated communicative practices from the immediacy of the interaction order to the mediated dynamics of sound recording that involved spatial and temporal separation between performers and audiences and a drastic reduction in the semiotic and durational affordances of the new medium. A secondary focus of my analysis, however, has been on the performers as they engaged in the work of remediation, compelled to rely on sound alone, engage audiences from whom they were separated in time and space, and accomplish their performances in the brief span of two and a half to four minutes. In the foregoing chapter on storytelling, for example, one of the key factors in accounting for the differences among the various efforts to adapt storytelling to commercial recording turned on the contrast between those narrators with prior experience as platform performers and those whose performance careers were bound up entirely with the new medium itself.

In this chapter and the next one, I want to bring the focus on the performer to the fore by looking in more depth at the recording careers of two pioneering phonograph performers, Cal Stewart and Charles Ross Taggart, whose professional personae, styles, and thematic emphases were in some ways closely similar but whose respective alignments to the new medium differed in important and illuminating respects.

We have already met Cal Stewart in earlier chapters, briefly in the chapter on burlesque sermons, more fully in the chapter on storytelling. Stewart was one of the earliest stars of commercial sound recording in the United States.[1] His recorded performances in his adopted persona of Uncle Josh Weathersby of Pumpkin Center, a fictional small town in rural New England, were immensely popular from the last years of the nineteenth century through his death in 1919 and for some years thereafter.[2] Throughout most of his recording career, as I suggested in chapter 3, Stewart was identified as a storyteller. The overwhelming majority of his recorded performances as Uncle Josh are in narrative form, but for illustrative purposes in launching this chapter, I focus on a single example that brings storytelling itself into relief, incorporating traditional tales from the classic American repertoire. I begin with an examination of the example itself and then broaden my scope to consider some of the larger dimensions of symbolic construction and cultural meaning surrounding Cal Stewart's career as "the talking machine storyteller" (C. Stewart 1903, 2).

"A Meeting of the Ananias Club" is a performative representation of traditional storytelling, a narrative account of a storytelling session in one of its most canonical American contexts. The country or village store was a privileged site of male sociability in rural and small-town American life, a place where the men of a community could gather to exchange the news, talk politics, comment on community life and the condition of the world more generally, and tell stories (Bauman 1972). Prominent in the expressive repertoire of these male gatherings were tall tales and other forms for the exploration of the epistemological tension between reality and fabrication, truth and falsehood, verisimilitude and exaggeration. The tall tale is a narrative genre that derives its interpretive effect by being framed as true but in which the circumstances of the narrated event are stretched by degrees to the point that they challenge or exceed the limits of credibility and rational understanding. Tall tales (and the metonymic condensations of tall tales that center on the core descriptive element) always make some claims on belief even as they exceed plausibility (Bauman 1986, 78–111; C. Brown 1987; Thomas 1977).

Cal Stewart, "A Meeting of the Ananias Club" (Victor 3103, https://purl .dlib.indiana.edu/iudl/media/415p295m43), Feb. 9, 1901

A meeting of the Ananias Club at Punkin Center, by Mr. Cal Stewart.
[Laughs]

Well, sometimes a lot of us old codgers,

we used t'get down t'Ezry Hoskins' grocery store
an' we'd sot aroun' 'n' eat cheese 'n' crackers or prunes,
or anything that Ezry happened t'have layin' around loose,
an' then we'd get t'spinnin' yarns.

Well, some o' the things we'd tell would just about put Ananias 'n' Safiry outa
     business if
     they was here now.
[Laughs]

One afternoon we was all sottin' around,
an' one feller said that down where he was born 'n' raised,
there was seven hills that sorta come t'gether,
an' ya didn't dare t'get out there 'n' talk any louder'n a whisper on account o'
     the echo.

An' one o' the summer boarders remarked that he warn't afraid to talk right
     out in fronta
     any lot o' hills that ever was created.
So he went out there an' he hollered as loud as he c'd holler
an' he started an echo a'goin'.

Well, it hit one hill, bounced off 'n' hit another,
over onto another hill,
gettin' bigger 'n' louder all the time till it got back where it started from,
hit a stone quarry,
knocked off a piece o' stone,
hit that feller'n the head,
'n' he didn't come to f'r over three hours.
[Laughs]

Well, we didn't say much fer a little bit till finally Jim Lawson remarked that
     he didn'
     know very much about echoes
but he calc'lated that he'd seen it rain just about as hard as anybody ever seen
     it rain.
An' somebody says, "Well, Jim, how hard did y'ever see it rain?"

"Well," Jim says, "one day last summer, I was a'settin' out on my porch an'
     it got
     rainin'.
Well, there was an' ol' cider barrel a'layin' out in the yard,
had both heads out of it,
bunghole up an' the water rained into that bunghole s'hard 'n' fast,

couldn' run outa both ends o' the barrel fast enough,
an' it swelled up 'n' busted.
[Laughs]

"Well, sir, we all nudged each other an' took a fresh chaw t'baccer,
an' finally old Silas Pettingill, he remarked that he'd never seen it rain
    very hard,
but he'd seen some mighty peculiar spells o'weather.
He allowed the coldest day he ever saw in 'is life was one day in August."

An' somebody said, "Well, how was that, Silas?"

He said, "That day, I was a'goin t'take m'mule down t'the blacksmith shop
    t'have 'im
        shoed,
an' I had 'im tied up to a pan o'popcorn out'n the yard.
Well, sun shone down on that popcorn s'hot that it got t'poppin',
'n' it flew around the old mule's ears,
an' he thought it was snow an' laid down'n froze t'death.
[Laughs]

"We disbanded after that.
We didn't feel like tempting providence any more at one settin'."
[Laughs]

"A Meeting of the Ananias Club" exploits two participation frameworks commonly found in American traditional storytelling. One framework is adapted to the performance of longer narratives during which the narrator holds the floor for an extended period. The performer may index the other participants by means of various metanarrational devices, but audience participation is limited essentially to back-channel responses. Stewart's recorded account of "A Meeting of the Ananias Club" is a narrative performance of this type. Cal Stewart is the sole speaker, from beginning to end. He does make one indexical gesture toward evocation of a co-present addressee, in his use of "Well, sir," after Jim Lawson's story: "sir" is a conventional appellative of respect. The overall performance, though, is an extended, fluent, virtuosic, but essentially monologic production. This framework, involving a solo performer, is clearly well suited to commercial recordings, and it is the format that Stewart employed virtually throughout his recording career. I should note that very few of Stewart's recorded narratives involve what folklorists would recognize as traditional folktales. In generic terms, the overwhelming majority of Stewart's recorded stories are narratives of personal experience or accounts of humorous or otherwise reportable occurrences in Pumpkin Center. "A Meeting of

Fig. 4.1. Advertising copy for Edison Phonograph dealers featuring a phonograph in a general store offering "Broadway vaudeville at Perkins Corners," *Edison Phonograph Monthly*, August 1908.

the Ananias Club" exemplifies the latter type, while incorporating thematic elements from the American repertoire of tall tales.[3]

The second participation framework is more interactive and conversational, involving more turn taking and collaborative coproduction of the performance. The narrative events in which this second framework prevails are characteristically sociable encounters; that is, they are occasions on which participants come together for the pleasure of each other's company and of the interaction itself, for its own sake. An especially significant feature of such traditional storytelling is that it is richly contextualized, indexing antecedent stories and storytelling events; the characters, roles, and relationships of participants; the emergent unfolding of the event itself; and the conventionalized frameworks for participation, relating to genre, ground rules for interaction, and other like metapragmatic elements.

An account of one such storytelling session at the country store serves Stewart as a frame tale for the representation of storytelling in its situational context in "A Meeting of the Ananias Club." In this recorded performance, Uncle Josh offers a detailed and culturally accurate representation of a sociable gathering at Ezra Hoskins's grocery store. The grocery store, as the principal venue for male sociability in Pumpkin Center, is the site of numerous such encounters that become reportable in Uncle Josh's narratives, bringing together the familiar cast of community residents that people Cal Stewart's recordings. Thus "we was all sottin' around Ezry Hoskins' grocery story," or some close variant thereof, serves as a contextualizing device in a number of Stewart's recorded performances.[4] In fact, Stewart was sometimes referred to in publicity materials as "The Corner Grocery Store Teller."[5] One of the most widely reproduced publicity photos of Cal Stewart depicts him on the porch of the "Punkin Center Generel [sic] Store," seated on a wooden cracker box with shavings (presumably from whittling) at his feet and a copy of the *Punkin Center Weekly Bugle* in his hand. Stewart's resort, across recordings, to the same settings and dramatis personae establishes a web of intertextual relationships among the recordings that serve as a broader contextualizing framework for the individual performances.

The recording opens with a general orientation, setting the scene for the action to follow, the telling of stories. It describes the cohort of Pumpkin Center codgers gathered at Ezra Hoskins's store, sitting around and snacking on groceries from his stock. "An' then," Uncle Josh tells us, "we'd get to spinning yarns." The event proceeds in a fashion typical of such sessions: a conversational topic—here echoes—serves as a stimulus and point of departure for a tall tale, and ultimately for a chained series of tall tales, all drawn from the traditional American repertoire. Hence the title "A Meeting of the Ananias Club": Ananias was a member of the church at Jerusalem who died immediately after uttering a lie; his wife, Sapphira, suffered the same fate (Acts 5:1–10). The Ananias Club is thus a classic liars' club. The first tale, an exaggerated account of the vocal constraints imposed by the echo effect of a set of hills, prompts a second one on the same topic, told by a summer boarder. The two echo stories set the tone and the narrative domain: extreme natural phenomena. Following the boarder's story, there is an interval of silence, a common feature of such yarn-spinning events (Welsch 1972, 11). Then Jim Lawson—familiar to us for his wiles as a horse trader—links his turn to the preceding one by acknowledging the topic of echoes, disclaiming further knowledge of it but offering a new topic construable as related, namely, torrential rain, another extreme natural phenomenon. Jim's suggestion that he knows something reportable about heavy rain serves as a tacit offer to relate what it is he knows, and with smooth uptake, one of the other participants asks him, "Well, Jim, how hard did y'ever see it rain?," amounting to acceptance of the offer. Jim then relates his whopper as an answer to the interlocutor's question. The other

Fig. 4.2. Cal Stewart as the General Store Storyteller, *Cal Stewart's Punkin-Center Weakly Bugle,* ca. 1917.

participants respond to Jim's whopper with what amounts to a positive evaluation in local terms: "we all nudged each other." The evaluation makes clear that these storytelling sessions at the store are performance events, occasions for the display of communicative virtuosity, and subject to evaluation by an audience.

The exchange revolving around Jim Lawson sets the pattern for the next round. Silas Pettingill ties his tale to Jim Lawson's in a manner similar to the way that Jim related his story to the summer boarder's: he offers topical discontinuity regarding rain but continuity regarding extreme weather, with a shift to extreme cold. Again, one of the participants accepts Pettingill's tacit offer to tell about "the coldest day he ever saw'n his life" by asking him about it, and Silas follows with his account of his mule confusing a shower of popcorn, set off by the August heat, for snow, and lying down to freeze to death. Silas's account about his mule freezing to death in August is the topper, serving as a warrant to bring the session—the meeting of the Ananias Club—to a close, in tacit acknowledgment that his tale is the most outrageous of the lot.

Stewart's representation of a storytelling session is impressively true to the ethnographic evidence. The sequencing and turn taking by which the event unfolds, including topical sequencing, intervals of silence, evaluative judgments of the performance, and termination all accord well with the accounts of such sessions as documented by folklorists and other firsthand observers (Bauman 1972; Thomas 1977; Welsch 1972). "A Meeting of the Ananias Club" is a meta-narrative, a story about a performance event in which traditional storytelling becomes an object: the *narrated* event is a *narrative* event that took place "one day last summer" at Ezra Hoskins's store.

Uncle Josh delivers the story in his characteristic speech register, as suggested in the preceding chapter: informal, vernacular, marked by "rube" dialect forms (such as deletion of final -*g* from -*ing* endings, the prefix *a*- attached to the present participle, as in "a-swimmin'," the preterit form *sot* for *sat* or *set, calculate* for *suppose, expect, intend*, and so on). While Uncle Josh's speech displayed a few features that would have been recognizable indicators of New England dialect to those familiar with the region, he relied for the most part on thematic elements to ground the narrative in place. His overall speech style was an amalgam of phonological, lexical, and grammatical elements of relatively broad regional distribution in American rural vernacular speech that listeners throughout the United States would have recognized as a generalized "country" or "rube" dialect, perhaps not precisely like the dialect of their own region but with enough familiar elements to do the indexical work of establishing Uncle Josh as a quintessentially rustic figure.

If we examine the formal devices that Uncle Josh employs in recounting the meeting of the Ananias Club, we find this his performance exhibits many of the common features of American storytelling—including tall tale—sessions, including prominently the following:

1. The use of "well" as an opening and subsequently as an episode marker.
2. The polysyndetic, additive chaining of narrative clauses.

3. The frequent use of quoted speech and taking on of voices as a means of characterizing dramatis personae and foregrounding clever, humorous, reportable talk.

4. Use of idiomatic intensifiers to amplify the remarkableness of the natural phenomena that are the focus of the tall tales, such as "mighty peculiar spells o'weather." This usage is entirely consistent with the keying of tall tales, which trade in exaggeration, hyperbole, and a general stretching of the truth.

5. Uncle Josh's signature laugh, an exuberant cackle that signals the narrator's own amusement at the narrated event and serves as a functional substitute for audience response and as an episode marker, setting off the successive turns at talk in the narrated event, the tall tale session at the general store. Stewart's early recordings as Uncle Josh were labeled "laughing stories" in Columbia's 1898 catalog, assimilating them to the popular genre of "laughing records." Uncle Josh's trademark laugh was an important stylistic feature of Stewart's recorded performances throughout his career.[6]

Our examination of "A Meeting of the Ananias Club," brief and summary though it may be, reveals Uncle Josh to us as a talented narrator, highly competent in the ways of American vernacular storytelling. He is fluent, funny, a master of conventional form, schooled in the traditional repertoire, adept at anchoring his stories in familiar discursive and situational contexts. He displays both a knowledge *of* traditional, vernacular storytelling and a knowledge of *how* to tell stories himself. We can easily see Uncle Josh as one of the stars of Pumpkin Center expressive life. But of course, Uncle Josh was not the star of anything. Pumpkin Center was a fiction, a representation of a small, rural New England town created by Cal Stewart. Uncle Josh was likewise a creation of Stewart's imagination and artistic skill, a performative simulacrum of a traditional storyteller.

There is some evidence to suggest that Cal Stewart, like other performers on early commercial recordings (Cogswell 1984, 38), was something of a star storyteller in traditional, face-to-face milieux. While we do not know much about his early life or his exposure to or experience in oral storytelling in traditional contexts, a brief autobiographical sketch, published in 1903, reels off an extensive list of work experiences that include well-documented venues of male sociability in which storytelling was a likely occurrence: riverboats, lumber camps, and railroads (C. Stewart 1903, 7–8). More explicit reference to his ability as a traditional storyteller and pride in his skill occurs in Stewart's (1903, 8) claim that

"[I] have been a traveling salesman (could spin as many yarns as any of them)."
Recalling his years as a railroad man headquartered in Decatur, Illinois, Stewart
casts himself as the star of the sociable gatherings in the railroad yard: "I was the
principal comedian in an old box car down in the yards that served as a waiting
room for the brakemen. I was leading man there during the last six years I was
in Decatur."[7] In what is perhaps a more objective assessment of his early talent,
a Decatur newspaper article of 1910 recalls, "When 'Happy Cal' lived in Decatur
he was known among his fellow employees for his inimitable wit and his never
failing good humor. . . . During the winter he used to entertain his comrades
with funny stories at the old East Decatur station as they sat around the stove."[8]
The internal evidence of his recorded performances certainly points convinc-
ingly to firsthand experience of traditional yarn spinning. By the evidence of
Stewart's recordings in which Uncle Josh told traditional stories or recounted
the storytelling of others, he was demonstrably familiar with and adept at the
forms and practices of the traditional storyteller's art.

But Cal Stewart, the talking machine storyteller, was a very different kind
of performer than Uncle Josh. While he might simulate address to a co-present
interlocutor or multiparty audience by the use of terms of address ("Well, sir,"
"you folks"), Stewart told his recorded stories not to a co-present audience (a
gathered, immediate public) but to a dispersed, distant, and mediated public, constituted by the distribution and circulation of his commercial recordings. Stewart, that is, created a constructed representation of the local star
performer, to be conveyed primarily to audiences distanced from the kind of
milieu he depicted in his recordings. In this effort, he was not unlike the folklor-
ist, though his motivations and goals and representational epistemology may
have been different. Stewart used his knowledge and ability of the traditional
star performer to achieve stardom in a different sort of social, cultural, and performance milieu. What can we discover of how he achieved national stardom
in the mass medium of commercial sound recording?

When Cal Stewart made his first reliably documentable Uncle Josh Weathersby recording in 1897, he was somewhere in the neighborhood of forty years
old,[9] already an experienced—though not widely known—stage performer
in both theatrical and platform events including vaudeville appearances in an
early prototype of his Uncle Josh persona. Stewart's biographers make much
of his experience as an understudy to the popular actor Denman Thompson,
who played the rural Yankee character Uncle Josh Whitcomb in his enormously
popular play *The Old Homestead* (Bryan 2002, 234; Gracyk 2000, 333; McNutt
2011, 41–46). The first notice of Stewart's budding career as a recording artist to
be published in *Phonoscope*, the trade journal devoted to the nascent recording

industry,[10] observes that "Cal Stewart has been before the public for the past twenty-five years, as a character comedian and monologue artist," underscoring his extensive experience as a stage performer. While the article identifies him by reference to his established identity as a performer who objectifies and enacts a Yankee character—"a delineator of New England character" and "a representative 'Yankee' comedian"—it goes on to observe that "his Uncle Josh Weathersby records have made a decided hit," calling attention to his new medium of performance and thus marking a critical turning point in his career. A half year later, in February 1899, a full-page advertisement in *Phonoscope* once again identifies Stewart as a "delineator" of "the New England character," but the master heading of the ad is "Cal Stewart, the Yankee Story Teller," followed by the question: "Have you in your collection any of the Uncle Josh Weathersby series of stories?"[11] Here is Stewart explicitly in the guise of storyteller, in the Yankee persona of Uncle Josh, but foregrounding the collectability of the records, as durable commoditized objects. Stewart's storytelling has migrated from oral tradition and co-present venues to the domain of mass-produced commodities, available for fetishized accumulation. The commercial appeal to accumulation was a recurrent theme in the promotion of Stewart's recordings, as, for example, in this 1909 blurb in the *Edison Phonograph Monthly*: "Thousands of Phonograph owners are acquiring a complete collection of the Uncle Josh Records, getting the new ones as fast as they appear."[12]

Stewart's fame and the popularity of the Uncle Josh recordings grew rapidly. By mid-1900, the editor of *Phonogram*, another trade journal, could plausibly maintain that "Everybody knows Cal or ought to," as "the author of the quaint talks known to all talking machine enthusiasts as the 'Uncle Josh series.'"[13] No longer, then, is Stewart identified in terms of his theatrical roles or platform performances. In two short years, his primary public identity had come to rest on his Uncle Josh recordings. And the grounding of Stewart's identity and celebrity as a performer in the phonograph and in records as his medium of performance remained a core element of his career for the remainder of his life. In newspaper advertisements, announcements, and accounts of his personal appearances, Cal Stewart is routinely characterized in relation to his phonographic celebrity. Here are some examples:

1. An article published on March 4, 1902, in the *Coshocton Daily Age*, of Coschocton, Ohio, says of Stewart, "He is a pioneer of the phonograph's early days and is known in all parts of the country."[14] Note that this is a mere five years after Stewart made his first known recording.

2. An announcement of a personal appearance in Fort Wayne, Indiana, later that year identifies Stewart as "the famous Yankee story teller, in his character sketch of Uncle Josh Weathersby, so well known to all lovers of the Gramophone."[15]

3. A 1913 notice from the *Lincoln Daily News*, of Lincoln, Nebraska, speaks of "Cal Stewart . . . who is known from coast to coast as the 'Uncle Josh' of the phonograph records," and is accompanied on the same page by an advertisement that identifies Stewart as the "man who talked in your phonograph."[16]

4. A 1916 ad for a vaudeville performance in Stephens Point, Wisconsin, identifies Cal Stewart as "the man who made millions laugh," "the Uncle Josh of phonograph fame," and "the man who made you Uncle Josh records for your phonograph."[17] A closely similar ad in the *Iowa Recorder*, of Greene, Iowa, suggests that these terms of identification were part of Stewart's stock publicity materials.[18] "The Uncle Josh of phonograph fame" occurs repeatedly in newspaper coverage of Stewart's appearances.

It is worth noting that there was a reciprocal relationship between the trajectory of Stewart's career and the evolution of sound recording as a popular medium of home entertainment. Not only did the development of commercial sound recording provide Stewart with the defining basis for the realization of his performances and his career, but he was recognized as an agent of the burgeoning of the medium itself. Recall, for example, the first newspaper notice from 1902, quoted just above. Here, a scant four years after his first known records were issued, Stewart could already be identified as a pioneer of the industry, with a national reputation. Late in his career, an ad for a personal appearance in Reno, Nevada, could claim that Stewart was "the man who made the phonograph famous."[19] Even allowing for the hyperbole of advertising and the vanity of self-promotion, the claim is not implausible. To summarize, then, Cal Stewart was clearly and decisively a man of the new communicative technology, whose celebrity was inextricably tied to the nascent mass medium of commercial sound recording. He was indeed "the talking machine storyteller."

At the very end of his career, as if to confirm how closely tied Stewart was to the medium that made him famous, he made a record about buying a Victrola (he was recording for Victor at the time) on which he proceeds to play one of his own Uncle Josh records, adding a reflexive flourish to his acquired persona as the talking machine storyteller. In *Uncle Josh Buys a Victrola*, issued in 1919, Josh brings the new talking machine home to Pumpkin Center from New York, the source of such modern marvels in the cultural cartography of his rural world.

The record itself is largely a promotional pitch for the new technology of home entertainment, portrayed as an instrument of sociability and "all the good music there was in the world." After playing a selection of musical records to appeal to the various residents of the community, Uncle Josh puts on a talking record of himself talking about the same cast of Pumpkin Center characters that has gathered in his "settin' room."

From "Uncle Josh Buys a Victrola" (Victor 18793, https://purl.dlib .indiana.edu/iudl/media/o89227tc67), May 1919

Well, sir,
there was one record what told about our town of Pumpkin Center.
It told about Jim Lawson an' me
an' Nancy,
Deacon Witherspoon,
Si Pettingill,
an' Hank Weaver,
Rube Hendricks,
Lige Willett,
an' about everyone in Pumpkin Center.
Fust one would laugh,
and then the other.
It all depended who the joke was on.
[Laughs]
Cindy Lawson got madder than a wet hen
when it told about her.
She said whoever did the talking on that record
was hard put for something to do.
Gosh, I never laughed more in my life at one time.
[Laughs]

The response of the various folk gathered around the Victrola is amused laughter when "the joke was on" someone else but a lack of amusement, we infer, when the joking spotlight turns to them. Cindy Lawson got really mad when the record "told about her." Not realizing, apparently, that it was her host himself that "did the talking on that record," she dismisses the performer as a layabout with nothing better to do. There is a subtle suggestion here, if Cindy Lawson is any indication, that the rural folks of Pumpkin Center don't quite get it, don't fully understand how the new medium works. Cal Stewart, from his vantage point as the talking machine storyteller, understands it very well, treating the technology, its structures of mediation, and his own recorded performances as objects and expecting his sophisticated listeners, those at home with the medium, to share his reflexive understanding. Unlike his neighbors, Uncle Josh can laugh at himself.

Notwithstanding the priority of sound recording in the making of Cal Stewart's celebrity, he did not at all give up live performance. Indeed, he capitalized on his enormous popularity and widespread recognition as a recording artist by giving performances and demonstrations all over the country, maintaining an active touring schedule to the very end of his career. These live performances provided opportunities for audiences who knew him first and primarily through his records to enjoy him in person. The recordings were the primary frame of reference; the live appearances were framed as secondary derivatives of the recorded performances. We take that relationship for granted today in the popular music industry in which the issuing of a new album is the occasioning basis for a live tour (Auslander 1999, 27, 64), but the career of Cal Stewart reveals it to us in its formative moment.

Because record-buying audiences knew Cal Stewart through his recordings, they experienced him as Uncle Josh's *voice*, as a performer to be *heard*. As one newspaper account frames this aural engagement, "We doubt if there was anybody in the house last night who would not have recognized Cal Stewart, the 'Uncle Josh of the Phonograph,' the minute he uttered his first sentence."[20] Advertisements for Stewart's shows, though, characteristically made a point of these live performances as occasions to both "see and hear" the popular performer, but they emphasize as well that this multisensory, unmediated engagement is the exception, a one-time, not-to-be-repeated opportunity. One ad proclaims, "You will never have but one chance to see and hear America's greatest rural story teller—the man who made you Uncle Josh records for your phonograph."[21] Another suggests to readers that the show being advertised is "a treat you will never repeat seeing and hearing."[22]

Cal Stewart's ever-growing popularity as a recording artist and his orientation to a broad commercial public brought about a gradual shift in his public persona. While he never completely abandoned or lost his identification with New England and his image as a Yankee comedian, the regional grounding of his character gave way increasingly to a broader and more diffuse identification with American rural or "country" life at large.[23] But from midcareer onward—say, after 1908—references to "New England" or "Yankee" fade away in favor of characterizations of Stewart as "unrivalled [in his] impersonation of country folk types," as "the best liked and best known rural comedian of the country," or as "without peer in his creation of country humorist," and even, ultimately, as "America's greatest rural story teller."[24]

When Cal Stewart embarked on his recording career at the turn of the twen-tieth century, rural life in America was the focus of intense scrutiny and critical assessment. This was a period of great change in the United States and of public preoccupation with change, heightened by the fin de siècle reflexivity char-acteristic of such transitional moments. The two decades that spanned Stew-art's active career as a recording artist represented a watershed period in the transformation of the American economy from predominantly agrarian to pre-dominantly industrial, with a concomitant trend toward the industrialization and scientific rationalization of agriculture, a burgeoning of consumerism, an acceleration in rural-to-urban migration, and a massive increase in immigration from eastern and southern Europe. All these factors challenged deeply rooted ideologies that located the founding essence of American republicanism in an agrarian way of life and identified the farmer as the quintessential American (Danbom 1979; Diner 1998; Hofstadter 1955). Within this broad arena of flux, the symbolic construction of rural America was an active and contested arena of cultural and ideological production as policy makers, reformers, economic en-trepreneurs, and, of course, rural people themselves attempted to comprehend and shape the past, present, and future of agrarian life (Bowers 1974; Brown 1995; Danbom 1979, 1995, 132–184; Diner 1998, 102–124; Rugh 2001, 181–182).

Uncle Josh's location in this cultural field is very complex, implicating an impressive array of features and factors. Constructed from its inception in the late eighteenth century as a vehicle for embodying social contrast and change, the Yankee rustic remained a durable symbolic figure into the early decades of the twentieth. In the latter period, however—that is, in Cal Stewart's heyday—the rustic stereotype appears to have become more polarized, tend-ing either toward benign nostalgia or burlesque ridicule.

Stewart's own portrayals of rural life fell on both sides of the ledger. On the one hand, his stories depict the people of Pumpkin Center as sociable (as in the gatherings at the general store), content with their way of life, and open to a hu-morous view of community social relations. At the same time, though, Stewart suggests in some of his recordings that his Pumpkin Center townsmen were not fully competent in the management of civil affairs, especially in the domain of governance. In *Political Meeting at Pumpkin Center* for example (Standard [Columbia] A287, mx. 1710, April 1904), he features a candidate for the state legislature as foolish in his framing of issues and inept at public speechmak-ing: "In my motions afore the legislature I intend to make will be one that calls for the aforesaid fact that each and every fire apparatus in this hyar state shall be examined each and every ten days before each and every fire." *Uncle Josh at a Meeting of the School House Directors* (Columbia A371, mx. 1504–10, recorded between 1903 and 1908) portrays the directors as bumbling stewards of the Pumpkin Center school: "I make a motion that we build a new school house outa the bricks of the old school house, an' that we don't tear down the old school house till the new one is built." These burlesque portrayals seem to corroborate the concerns of progressive reformers that the rural communities commonly held up as embodying the essence of the American democratic spirit had become so thinned out in what Horace Kallen (1998, 83) called "the degenerate farming stock of New England" as to no longer to be capable of sustaining a vigorous democratic polity in the age of modern, complex society (Bailey 1915, 108; Lippman 1922).

Notwithstanding the element of critique and ridicule that surfaces in some of his performances, however, when it came to the marketing of Uncle Josh as an exemplar of country life, all was sunshine and nostalgia. Stewart himself, as well as his record companies, cast Uncle Josh increasingly in a nostalgic and sentimental key. "In Uncle Josh," he suggested, "you have a . . . character chuck full of sunshine and rural simplicity." Indeed, the key throughout Stewart's career was nostalgic retrospection: notices and advertisements in newspapers and trade journals characterized Stewart's manner as "quaint," Pumpkin Center as "romantic," and the scenes depicted in the recorded stories as "old fashioned."[25] But overarching all of Uncle Josh's various alignments vis-à-vis the life of his time is his primary identity as storyteller. In Cal Stewart's construction of Uncle Josh, storytelling comes to the fore: it is emblematic of "country" life, its preeminent expressive form.

And here, I would argue, lies the primary significance of Cal Stewart's re-mediated stardom, from "leading man" in a sociable group of railroad men to national celebrity as the talking machine storyteller: the promulgation of a popular image of storytelling as residual culture, a form of expression that was fated to decline with the social and cultural formations in which it was rooted. To be sure, by the time Cal Stewart came along, the casting of story-telling as a key element in a vanishing rural way of life was a long-established element of modern social theory, dating back at least to the late seventeenth century (Bauman and Briggs 2003). But the writings of social theorists were produced by intellectuals for intellectuals and were based largely on the advent of modernity in Western Europe. Stewart, however, addressed a large popular audience, in terms that resonated with their own American experience—or, more accurately, with the experience of that burgeoning population of urban, middle-class Americans.

During the first two decades of the twentieth century, the phonograph was a part of an emergent consumer culture, targeted at the urban bourgeoisie and the growing class of prosperous, mechanized farmers with sufficient disposable income and leisure to expend on home entertainment. We may gain a sense of the demographic reach of Stewart's appeal from reminiscences of the period. Consider, for example, the report of Jim Walsh (1951b, 22), one of the principal authorities on Stewart's career, that "a friend of mine in Decatur, Illinois, has told me that a certain outlying area there is known as 'Punkin Center,' because of the resemblance of its farm types to those in the Stewart records." Walsh's friend is reporting the view from the urban center, looking outward beyond the city line to an "outlying area," a rustic hinterland. It seems more than likely that a similar impulse would account for the several dozen other Pumpkin Centers (or Punkin Centers) that dot the American map, including two in my own state of Indiana (McNutt 2011, 155). Another observer, from California, recalled, "We are reminded of a time we stood at a talking machine booth at a state fair and noticed that the farmers who paused invariably asked for 'Uncle Josh' selections. 'Uncle Josh' was manufactured in the city. He was the city man's idea of a farmer, a creation which the real farmer accepted as being exquisitely funny because it was so far removed from his own experience."[26] The state fair was a display event for modern, up-to-date farmers committed to the scientific improvement of agriculture, precisely the kind of farmers who considered themselves far removed from the old-fashioned hicks of Pumpkin Center. Other country people found the humor of performers like Cal Stewart less amusing. Respondents to a national survey conducted by the secretary of agriculture in 1913–1914 deplored the stereotypes that "rube" humorists purveyed to the general public. A farm woman from New York wrote, "The thing that seems to me to most need remedying is the attitude most town people have toward the farmer. He is represented either as a 'Rube' with chin whiskers and his trousers in his boots or as having several motor cars bought with his ill-gotten gains from farm products figured at the highest retail prices. One of these ideas is just as inaccurate as the other" (USDA Office of the Secretary 1915, 24). A farmer from Ohio voiced a similar charge: "The farm folks are treated with contempt and ridicule. Scarcely a daily paper or periodical of any kind by caricatures and pictures the farmer as old 'Hayseed' with a make-up that is disrespectful and not true" (1915, 24).

Now, consider the image of the storyteller that Cal Stewart offered to the audiences that listened to his records, a constellation of tightly integrated features that constructed the storyteller as an anachronism. I emphasize the element of *construction*. The aggregate image of the storyteller that emerges from what follows is, like all social stereotypes, grounded only partially and diffusely in empirical reality (whatever that is), and virtually every element of the stereotype has been qualified and nuanced by much critical research. And again, like all stereotypes, it is an ideological construction: positioned, interested, differentially valued, contested. Nevertheless, the principles and tenets that entered into the construction of the storyteller as a relic, both by direct attribution and by contrast with its ideological opposites, were widely held in the first two decades of the twentieth century and served as guiding principles for policy and practice.

First, Uncle Josh is a New Englander, though the Yankee component of his persona diminished—but never fully disappeared—over the course of his career. Rural New England, in Stewart's time, was widely recognized as a region in economic and demographic decline, its farms unsustainable in the modern agricultural economy and its population diminishing as the young people decamped for the city (Brown 1995; Wood 1997). The venue for Uncle Josh's storytelling is the general store, an institution also in decline, in the face of new models of retail commerce, such as mail order, chain stores, or urban department stores (Marler 2003; Trachtenberg [1982] 2007, 130–135). The general store was the quintessential site of a form of male sociability in which storytelling was a privileged mode of communication. That form of sociability, in turn, was deeply dependent on the rhythms of rural life and upon the *gemeinschaftliche* social relations in which community members were closely familiar with each other's lives in common. Urban life was supposed to be different: mobile, rationalized, impersonal.

The storyteller himself, embodied in Uncle Josh, is an old codger, marked by manner and name—Uncle Josh—as belonging to a senior generation and to extended family relations; he is the uncle you left behind in the country when you moved to the city and drew your web of kinship relations inward to the nuclear family, or the fictive "uncle" who was a respected elder back in your small town. He delivers his stories in a nonstandard, vernacular dialect, dramatically contrastive with the standard English taught in the schools and held up as a requirement for social and economic advancement. His speech style is colorful, heightened, idiomatic, and personally distinctive, all expressive features that were likewise flattened out by standardization and disvalued by language ideologies that privileged unadorned, clear, logical, objective, propositional discourse over artistic, highly figurative, allusive, and personalized narrative.[27]

What is especially compelling about Stewart's stardom, of course, given that it was so centrally grounded in his figuration of the old-fashioned country storyteller, is that it was equally strongly dependent on the new medium of phonographic sound recording. New communicative technologies, such as the telephone and the phonograph, were also widely understood in Stewart's day, as now, to have a transformative effect on social life, introducing mediated, distantiated, and commoditized modes of sociality that contrasted with and, to a degree, displaced the immediacy of co-present, socially embedded, spoken interaction. Cal Stewart's star persona, the talking machine storyteller, is thus a hybrid construction: both an ultramodern creation and a relic. He relied for his storytelling on one of the very communicative technologies fated to render his favored mode of expression obsolete. But a hybrid construction is well suited to a state and a period of transition, as remediation is a necessary element of nostalgia. Stewart's media stardom prefigures the far better-known rube characters of radio, film, and television, but none of them makes storytelling so central a part of the rube persona or goes so far in casting storytelling in the popular imagination as an outmoded relic of our social and cultural past. Seeming on the face of it to preserve traditional storytelling and, in the virtuosic performances of "America's best loved storyteller," to extend its reach to mass audiences, the remediation of storytelling on Cal Stewart's recordings served to intensify the conditions of its decline by casting it in the popular imagination as residual culture—at best an element of nostalgia, at worst a corny index of a historical past best left behind.

NOTES

1. For biographical information on Cal Stewart, see Bryan 2002; Feaster 2006; Gracyk 2000, 332–338; McNutt 2011; Petty 1974; Walsh 1951a, 1951b, 1951c, 1951d.

2. Patrick Feaster (pers. comm.) points out that "at first, Stewart often presented his rube character not as a New Englander but as a New Jerseyite. He played an 'Original Jersey Farmer' . . . on the stage at Fort Wayne in 1892, and even some of his early recordings give Josh's New Jersey origin." Nevertheless, the New England setting of Pumpkin Center and Uncle Josh's persona as a Yankee storyteller became solidified very early in his recording career and remained in play thereafter.

3. Type 1920 Contest in lying (Baughman 1966, 59); Motifs N520 Lies about mountains and hills (Baughman 1966, 547); X1764 Absurd disregard of the nature of echoes (Baughman 1966, 590); X1764(b) Echo sounds a long time after sound is made (Baughman 1966, 590); X1654.3.1(a) In hard rain, the rain goes into bunghole of barrel faster than it can run out both ends (Baughman 1966, 575); X1632.3*(a) Heat causes kernels of corn to pop, causing animal to freeze to death (Baughman 1966, 569); X1633.1 Heat causes corn to pop in crib or in field. Animals (cows, horses, mules) think the popping corn is snow, freeze to death (Baughman 1955, 570).

4. See, e.g., "Jim Lawson's Hoss Trade," discussed in chapter 3 (https://purl.dlib.indiana .edu/iudl/media/534f661toq) and "War Talk at Pumpkin Center" (Victor 17820, matrix B-16102; https://purl.dlib.indiana.edu/iudl/media/544b693v59), June 14, 1915.

5. See, for example, *Daily Northwestern* (Oshkosh, WI), June 23, 1908.

6. The features and significance of laughing records in the early commercial record catalogs, including those of Cal Stewart, is discussed in illuminating detail in J. Smith 2008, 15–49.

7. *Daily Review* (Decatur, IL), December 14, 1902, 13; cited in Feaster 2006, 214 n.

8. *Daily Review* (Decatur, IL), April 8, 1910, 7; cited in Feaster 2006, 214 n.

9. Patrick Feaster, who has done the most exacting research on the matter, estimates that Stewart was born about 1860 (pers. comm.).

10. "Our Tattler" 1898a, 12–15.

11. *Phonoscope* 3:2 (1899): 6.

12. EPM, April 1909, 19.

13. *Phonogram* 2:1 (1900): 30.

14. *Coshocton Daily Age* (Coshocton, OH), March 4, 1902, 1.

15. *Fort Wayne (IN) Sentinel*, November 27, 1902, n.p.

16. *Lincoln (NE) Daily News*, September 20, 1913, 3.

17. *Stephen's Point (WI) Daily Journal*, October 11, 1916, 6.

18. *Iowa Recorder* (Greene, IA), January 31, 1917, 5.

19. *Reno Evening Gazette*, December 2, 1918, 8.

20. *Iowa City Citizen*, May 12, 1917, 3.

21. *Stephen's Point Daily Journal*, October 11, 1916, 6.

22. *Elyria (OH) Chronicle*, April 19, 1915, 4.

23. McNutt (2011, 5, 18–19, 67) makes a note of this shift at several points in his biography of Stewart.

24. EPM, January 1909, 19; *Iowa City Citizen*, May 10, 1917, 5; *Iowa City Citizen*, May 12, 1917, 3; *Steven's Point Daily Journal*, October 11, 1916, 6.

25. EPM, September 1908, 26; *Coshocton Daily Age*, March 4, 1902, 1; EPM, September 1908, 26; EPM, January 1909, 18.

26. *Oakland Tribune*, April 2, 1926, 24; cited in Feaster 2006, 247 n.

27. This ideological tension has a long history. See Bauman and Briggs 2003; and Silverstein 1996.

FIVE

"SOMEBODY STOLE MY TUNE!"

Charles Ross Taggart and
Country Communicability

INTRODUCTION

While Cal Stewart was preeminent among early recording artists as a symbolic embodiment of the rural storyteller, he was not without fellows in the record company catalogs of the day. The presence of other rube performers on early commercial recordings attests to the popularity of the country talker as a performable figure, not only on records but in the popular entertainments like vaudeville, medicine shows, and lyceum programs from which the early record companies drew many of their performers, themes, and genres. Among those early rube performers, second only to Stewart in popularity and prominence in the record catalogs, was Charles Ross Taggart, whose performance persona as the Man from Vermont and performed representations of New England rural life were closely parallel to Stewart's own. But Taggart added a second element to his role as a rural character: in addition to enacting the Man from Vermont, he performed under the sobriquet of the Old Country Fiddler, and his record titles foregrounded that identity. Traditional fiddling is readily recognizable as an emblem of country life in contemporary popular culture, but Taggart was the first recording artist to build it into his performance personae and his portrayals of country life. He is certainly a storyteller in the mold of Cal Stewart, but his recorded narratives incorporate fiddle playing as a stylistic and thematic element that endowed his records with a distinctive character during the formative period of sound recording as a medium of home entertainment. Moreover, Taggart's engagement with commercial recording and his alignment to the new medium contrasted with Stewart's in illuminating ways, and his portrayal of country life was in many ways richer than Stewart's in its construction of the forces of change affecting rural life in the early decades of the twentieth century. In this chapter, developing further on themes introduced at the end of chapter 3, I examine Taggart's career, his engagement with the phonograph, and the nature of his recorded performances as the Man from Vermont and the Old Country Fiddler.

One aspect of Taggart's recorded figurations of the Man from Vermont that is especially nuanced is his depiction of the communicative world that he inhabits, both in his home community in rural Vermont and in his ventures out into the wider world, what I will call, following Charles Briggs, *country communicability*. In Briggs's formulation, "communicability refers to socially situated constructions of communicative processes—ways in which people imagine the production, circulation, and reception of discourse" (Briggs 2007a, 556; see also Briggs 2007b). That is to say, communicability is, first of all, a cultural and ideological construction, a way of conceiving, interpreting, and evaluating how the social world is communicatively organized. Importantly, the notion of communicability demands attention to the "situated" nature of such constructions, the contexts within which systems of communicability are articulated and the sites and trajectories of discursive production, reception, and circulation that map how communicative processes are spatially and temporally distributed and organized. That is to say, systems of communicability are chronotopic. And finally, like all cultural constructions, systems of communicability are themselves discursively constituted, actualized in specific discursive forms and practices. Communicability is thus an inherently reflexive concept, focusing attention on communication about communication, discursive events and processes and forms that make claims—explicit or implicit—about discourse.

Briggs and I have suggested in a number of works (Bauman and Briggs 1990, 2003) that performance, which tends to be among the most reflexively heightened, memorable, repeatable, circulable, and hence shareable sectors of any communicative economy, represents a highly productive domain for the investigation of communicability. The constructions of communicability I treat in this chapter, as represented on Taggart's Old Country Fiddler recordings, are *constituted* by performance: symbolically encoded, enacted, placed on display before an audience. The performances I analyze are constructions of communicability in two principal senses. First, the Old Country Fiddler, as Taggart enacts him, is an *embodiment* of country communicability. He is, in Asif Agha's (2007b, 177) phrase, a characterological figure, a stereotypical, performable exemplar of the kind of person linked to particular registers and, for our purposes, to particular chronotopes. A characterological figure is the vehicle by which a chronotope takes on flesh and agency and becomes artistically visible (cf. Bakhtin 1981, 84; see also Blommaert 2015).

At the same time, the characterological figure of the Old Country Fiddler is himself a performer, a storyteller who creates narrative representations of the social world in which country communicability has its characteristic place. The dramatis personae, communicative forms and practices, situational contexts, social roles, structures of participation, and systems of affect and the like that he portrays in his narrative accounts of life in Pineville and other spaces through which he moves make up a composite picture of country communicability as a cultural and ideological system in a vernacular key. In his performances, the Old Country Fiddler embodies, recounts, and experiences country communicability, mediated through sound recordings. Commercial phonograph recordings represented the first of the acoustic mass media and served as important vehicles for the expression and dissemination of country communicability. The communicable ground of Charles Ross Taggart's recorded performances centers on rural Vermont in the early decades of the twentieth century and by indexical extension on rural New England more generally. As the discussion to follow makes clear, however, Taggart's performances counterpose rural communicability to contrastive—principally urban—communicable formations. Systems of communicability are always multiple, open to competition, refraction, or deformation by other formations in the communicative environment.

CHARLES ROSS TAGGART

Charles Ross Taggart was born in Washington, DC, in 1871, but grew up on his grandparents' farm in Topsham, Vermont, to which his widowed mother had returned when her son was two years old.[1] Taggart showed an early aptitude for music and studied first with itinerant music teachers; then with a teacher in Montpelier, to which he commuted twice a week; and later at the New England Conservatory in Boston, where he pursued piano and voice. He bolstered his skills in verbal performance at the Emerson School of Oratory, where the curriculum included voice culture, gesture, dramatic interpretation, and rhetoric. He was in all respects a highly trained verbal and musical performer.

Returning to Topsham from his studies in Boston, Taggart found himself at loose ends, trying his hand at farming, teaching music, teaching at the "deestrict school," clerking, and "tinkering watches," but feeling that he "was not progressing very rapidly either in a professional or business way." A fortuitous encounter with Frank G. Reynolds, a traveling entertainer from Boston, led him to consider a career in which he could put his "various acquirements in the speaking and musical lines" (Greer 1927, 84) to professional use, and he made his platform debut at the Topsham Town Hall in the fall of 1895. The performance was a

Fig. 5.1.
Redpath-Slayton
Lyceum Bureau
flyer for Charles R.
Taggart, entertainer,
the Man from
Vermont, n.d.

success, at least by Taggart's standard, yielding the grand profit of "seven dollars and a half," and Taggart's career as an entertainer was launched. He went on to sign with a series of regional lyceum bureaus and traveled the platform performance circuit as the Man from Vermont and the Old Country Fiddler, eventually becoming affiliated for more than twenty years with the Redpath Bureau of Chicago, the largest and most prominent lyceum and Chautauqua bureau in the country. In Taggart's view, Redpath could furnish him "the class of people to entertain" that he aspired to: "an audience of bright keen intelligent people that respect something good."[2] Tent Chautauqua—and with it, Taggart's career—began to decline in the late 1920s, but Taggart continued to tour until the effects of a stroke compelled him to retire from active performance in 1938. He died in Maine in 1953 at the age of eighty-two.

Taggart constructed his platform performances as the Man from Vermont and the Old Country Fiddler out of a variety of elements: fiddle tunes, piano pieces, novelty fiddling (e.g., "fiddle mimicry"), ventriloquism, recitations, the personation of a variety of rural characters, and humorous monologues. An advertisement in a lyceum trade journal suggests how Taggart billed his platform act: "He has scored again and again as a musician, an impersonator, an unequaled story-teller and as a 'fiddling wizard.'"[3] Publicity notices and reviews in local newspapers where he performed, however, tended to zero in on his fiddling and storytelling, as in a 1922 review that lauds his "marvelous 'fiddling'" and "inimitable story telling."[4] Often, though, his storytelling came to the fore. "He is reported as being especially clever in his quaint stories of the 'old town folks' of New England," reads one publicity announcement.[5] While Taggart's twin performance personae are thoroughly intertwined, we might suggest that the Man from Vermont highlights his role as a storyteller and his focus on talk while the Old Country Fiddler foregrounds his musical performances and involvement in the musical culture of his community.

A specimen program, included in a Redpath publicity sheet for "Charles R. Taggart, the Man from Vermont," is headed "Pineville People," headlining the fictional Vermont community that was the cartographic center of Taggart's performances. Among the Pineville folks listed are "Sile Haskins, the Village Storekeeper," "Sandy Laird," and "Uncle Zed Jackson, the Fiddling Shoemaker."[6] When it came to his recordings, however, the Old Country Fiddler (sometimes identified by name as Uncle Zed) played the central role and predominated in the titles of his records. In this role, Taggart played his fiddle and recounted stories about life in Pineville, the foibles and escapades of his fellow townsmen, and his occasional encounters with urban life on his visits to Boston or New York. From 1914 to 1924, Taggart recorded a number of his Old Country Fiddler routines for Victor, Columbia, and Edison, of which twenty-five different performance routines were issued plus a number of re-recordings of earlier releases.[7] Victor categorized his recorded performances as "Rural Monologue with Violin Specialty," foregrounding the stories but acknowledging the special contribution of fiddling as well.[8] Taggart suggests how the two performance modes are interrelated on his recordings in his conclusion to "The Old Country Fiddler in New York" (discussed at greater length below): "Well, I'll play ye 'Pop Goes the Weasel.' I get t'talkin', tellin' stories, I forget what I'm doin'." While we cannot know the precise relationship between Taggart's performances onstage and on record, it is clear that the stage allowed for more relaxed and extended presentation of his sketches. In a letter to one of his contacts at the Redpath Bureau, he notes, "I am busy boiling down my material into time limits of 3 minutes and 25 seconds."[9]

Where Cal Stewart took Uncle Josh onto the stage to boost his record sales and his already established fame as the talking machine storyteller, Taggart's trajectory was in the opposite direction: he perceived his recorded performances primarily as a means of augmenting his reputation as a platform entertainer. He encouraged the record companies on the one hand and the local Chautauqua committees on the other to make special efforts to advertise and distribute his records in the towns and cities in which he was scheduled to appear on stage.[10] Writing to one of his contacts at the Redpath Bureau, Taggart was explicit about his motivation for making records: "I make this attempt to boost 'The Man from Vermont' in the Lyceum world."[11] He suggested that the cities and towns where Victrolas were sold should be prime booking sites for his shows and envisioned that the recordings would enhance the publicity for his live appearances. Despite giving priority to his platform performances, Taggart appears nevertheless to have felt that his material was well suited to recording, as he suggested in a letter to one of his Redpath connections: "People are more pleased with what they *hear* than with what they see in my work."[12]

Redpath quickly got the point. They produced new publicity circulars that featured the availability of his recordings within weeks of Taggart's first recording session and included mention of the records in their notices in the *Lyceum News*, the leading trade journal for platform performers.[13] When Taggart later had a publicity photo made as the Old Country Fiddler listening to himself on a Victrola, Redpath used it in its *Lyceum News* advertising and in a new circular.[14] Victor, for its part, was quite ready to cooperate in cross-marketing Taggart's records and live performances. They sent circulars listing his records to each town in which he was scheduled to appear and contacted the record dealers in those venues to urge them to work with local sponsors "in advertising the Charles Ross Taggart records and the Chautauqua."[15] In short order, Taggart was able to report back to Redpath, "It is working."[16] There is no doubt that the popularity of Taggart's records upped his stock with Redpath. By the end of 1916, not quite two years after Taggart made his first recordings, one of the bureau executives noted in a bulletin to all the Redpath managers, "He certainly made it mighty big this year and his reputation is increasing right along through his Victor records. They are constantly increasing his price per record and now they are paying him a bonus to make records exclusively for them as the Columbia people are anxious to make records of this kind of entertainment."[17] For the local sponsors of his appearances, the popularity of Taggart's recordings seems to have paid off as well in increased receipts. In one of his regular reports back to Redpath in early 1918, he was eager to inform them, "I find my Victor work helps amazingly in my independent [i.e., lyceum] work. The committee at Burlington, VT cleared $50.00 on my entertainment there last Thursday."[18]

It is difficult to discover how early commercial recordings were received by members of the consuming audience. There is a suggestive passage in the Victor promotional copy advertising Taggart's second recording, issued in June of 1915: "The two Taggart selections issued in March made a real hit with the Victor public, as many comments show. Says a Vermont customer: 'We hope you will decide to give us some more of those records by Charles Ross Taggart. He knows just how to picture the old-time Vermont Yankee without overdoing it, and we have greatly enjoyed his record. I am a Vermont Yankee myself, and as we realize that this old type is rapidly passing away, it seems to me that the time may come when these records will have a historic value.'"[19]

Whether this is a genuine comment made by a true Victor customer or a promotional fiction authored by a Victor publicist, it suggests one possible interpretive positioning of Taggart's Man from Vermont persona. The stereotypical Vermont Yankee that Taggart enacts is, as I have suggested, a characterological figure. As Taggart performed him, the Man from Vermont is an "*old-time* Vermont Yankee," an "*old* type," representing a fading past that is "rapidly passing away." The putative customer explicitly historicizes Taggart's character but with a touch of nostalgic regret that this social relic of a bygone era is marked for extinction. The Yankee, of course, had deep roots as an expressive resource, reaching back to the colonial era, but in the course of his long career in the American cultural imagination, he underwent an aging process, from energetic and exuberant—if rough—symbol of the frontier to aging denizen of a played-out region. In Taggart's recordings, we find him cast as a codger in the waning days of his symbolic life course.[20] This interpretive and ideological construction is certainly consistent with one widely current public image of the Vermonter in the late nineteenth and early twentieth centuries as caught in a time warp. A 1927 *National Geographic* portrait of "The Green Mountain State" casts Vermont as "one of the most truly American of our States. Its people have hardly changed in their essential elements in a century" (quoted in B. Harrison 2005, 480). This observation cuts two ways. On the one hand, it gives a nod to Vermont and its (White) people as quintessentially American, personifications of the authentic American character. At the same time, though, it marks the state as conservative in its essence, a relic area. To remain unchanged, of course, is to be left behind in the face of rapid and radical change in the broader society. In the early decades of the twentieth century, Vermont was thus a chronotope, a cultural model that fuses particular orders of temporality and spatiality and, in turn, gives shape to a particular genre of texts or performances. What I want to explore in more detail in this chapter is just how Taggart's recorded performances construct the stereotype of the "old-time Vermont Yankee" with which the author of the letter in the Victor ad identifies so strongly. What role does fiddling play in his recordings? And how do Taggart's records index the social change that exists in symbolic tension with "old-time Vermont Yankee" social life?

DIALECT

One of the first things that strikes the listener on hearing Taggart's recordings is his conspicuous dialect, a discourse register that marks him unmistakably as a New Englander.[21] A discourse register is a conventionalized repertoire of co-occurrent performable signs that index particular kinds of persons, communicative practices, and situational contexts. In their promotional materials for Taggart's recordings, the record companies emphasized the authenticity of his regional dialect, grounded in his true New England origins. "Mr. Taggart is a real 'Yankee,'" a Victor promotional flier proclaims, "and doesn't have to 'put on' any so-called dialects."[22] He is not merely a stage Yankee but a real one, "pure and undefiled," as one interviewer puts it. "His dialect is not assumed" (Greer 1927, 86; cf. Gal and Irvine 2019, 229–231).

Taggart, as noted, grew up in Topsham, on the eastern side of Vermont, in what is recognized by dialect geographers as the eastern New England dialect area, but his dialect—at least in its phonological aspects—inclines more toward western New England pronunciations. He tends to pronounce postvocalic /r/, though with occasional variation, as in "star" /sta:/ or "there" /ðɛə/. Likewise, /a/ and /ɔ/ are for the most part distinct in his speech, though they are occasionally merged, as in "washed" /wɔʃt/ or "comets" /kɔmɪts/. Grammatically, Taggart's dialect is marked principally in past-tense verb forms, some more common in eastern than western New England (e.g., "driv" for "drove," "ketched" for "caught," "riz" for "rose"), others widely current in New England (e.g., "sot" for "sat," "et" for "ate"). The same pattern holds true for lexical items, with "whiffletree" for "singletree" and "tie-up" for "cow-barn" or "sass" for "sauce," more common in eastern New England. Taggart seems especially fond of "buttery" for "pantry," which he uses in several routines as a distinctively New England term, and his speech is conspicuously flavored with colorful, regionally common interjections, like "I swan to man" and "Gosh all hemlock."

It is notable, in more general terms, that many of Taggart's dialect features, especially lexical and grammatical forms, are described by informants for the *Linguistic Atlas of New England* as "older" or "old-fashioned." The fieldwork for the atlas was carried out between September 1931 and October 1933, only a decade and a half or so after Taggart's earliest recordings, which suggests a deliberate effort on his part to select for dialect forms that mark his speech as old-timey and quaint. The effect is reinforced by the high degree of contraction and elision that characterizes Taggart's speech as casual, vernacular, and nonstandard. "There's a good'eal t'everything," he says, summing up one of his recorded performances. He is especially given to syncope, as in such regionally common forms as "prof'table," "butt'ry," or "cal'late," and to aphaeresis, as in "'fore," "'merican," "'tention" (Fisher and McDavid 1973). All in all, Taggart's speech constitutes a significant element of his performance personae: the Man from Vermont and the Old Country Fiddler. It marks him as an old-fashioned, rural New Englander, his casual, vernacular speech perfectly consistent with the country character he represents.

MALE SOCIABILITY AND PINEVILLE INTELLECTUAL LIFE

We are already familiar, from our visits to Cal Stewart's Pumpkin Center, with one of the most emblematic locations in the communicable cartography of rube communities, so let's start in the corresponding site in the Old Country Fiddler's community of Pineville, at Sile Haskins's store. Haskins is one of the most prominent of the recurrent dramatis personae in Taggart's routines besides Uncle Zed himself. The general store, as we have seen, was the center of male sociability in American rural communities, the place where men gathered to talk, tell stories, discuss the news of the day, share the goings-on in their locale, and, as we shall see, ponder the mysteries of the universe (cf. Tandy 1964). Sessions at the store were at—or very near—the center of the rube-record chronotope. Taggart offers us a nuanced and revealing view of a sociable gathering at Haskins's store in Pineville in "The Old Country Fiddler on Astronomy."

Charles Ross Taggart, "The Old Country Fiddler on Astronomy" (Victor 18148; https://purl.dlib.indiana.edu/iudl/media/d56z908r6n), June 22, 1916

We got t'talkin' one evenin' over at the store, about the sun and the moon. Y'see they had the subject over t'the lyceum one evenin': "resolved that the sun does more good'n the moon." And they thrashed it out there till nigh onta ten o'clock. And so some o'the men there in the store took sides one way and some another.

Sile Haskins said of course the sun did more good'n the moon, 'cuz the sun was in the center of everything. He said that the world, an' the moon, an' the stars, an' comets, and meteors, an' the who:le business all sailed round the sun every day. An' he got'n egg an' told Ab Tolliver t'hold it, an' he took a lemon in one hand, an' a potater in the other, an' showed 'em just how the universe operated.

Eh, but Ed Judson, he said that any fool could see with 'is own eyes that the sun riz in the east an' sot'n the west, but where in thunder it went to in the night, he didn' know. Might come back over, ready t'start again the next mornin', or t'might go round on t'other side.

Jock Sullivan said he'd bet a dollar his boy c'ld figger it out by algebry.

Eben Button said if the earth got t'movin' round, and tippin' up edgeways, t'would jiggle the Atlantic Ocean, so it'd slop all over everything, like enough puttin' the sun out altogether, an' the moon'n the stars would like enough be washed away.

Just then, Jim Sanders come in an' wanteda know what all the talk was about. I told 'im that we was a'continuin' the discussion that they had over t'the lyceum as to which did the most good, the sun or the moon. And he said of course the moon did more good'n the sun, for the moon shone'n the night, when it was dark, and the sun shone right'n the daytime, when y'didn't need it.

Then Garl Maston spoke out, an' Garl said warn't none of 'em got it right. He said he seen'n an astronomy book once, as how most o' the stars was twice as big as the sun, only they was quite a ways off, an' the sun was considerable bigger'n the whole world.

Just then, Pete Pepper jumped up an' said, "Well, gosh all hemlock! What does Pineville amount to anyway, I'd like t'know? Why, if this world, should, eh, get lost, there wouldn't nobody notice it. Heavens an' earth!" says Pete. "I'm goin' home. Good night!"

But just then Elder Bowker, the Baptist minister, come in after's mail, an' Sile called 'im over, an' asked 'im how t'was 'bout this sun moon business, an' he straightened it out for us, well's 'e could.

But land sakes, we did feel pretty small. Least I did. T'think of all them planets, an' solar systems, an' zodiacs, an' meridens, an' constellations, an' what seemed the hardest t'see through was there warn't no end t'anything. When y'got clean out t'th'edge, you just begun.

All the men was pretty quiet, when we went home that night.

Now, I tell ye, I couldn't help thinkin' that every one 'un us was just like that man in Pollok's *Course o' Time* that I used t'read in school:

"A man there was,
and many other such
y'mighta met,
who never had a dozen thoughts
in all 'is life,
but told them o'er,
each in its customed place
from morn till night,
from youth till hoary age."

Well, I guess it's somethin' like old Sylvester Boynton used ta say, "There's a good'eal t' everything."

As a social center and perhaps the most prominent feature of cartographies of country communication, the general store has been a ready expressive resource. Accordingly, the store has been a ready expressive resource for performance, a setting for the representation of country ways of speaking, usually in a humorous key. Cal Stewart's "Meeting of the Ananias Club" has already introduced us to the pleasures of this colorful setting. A characteristic feature of evening gatherings at the general store is the competitive, even agonistic tenor of the interaction, as participants vie with each other in verbal self-aggrandizement, whether by claiming superior knowledge, wider experience, or deeper expertise, zinging each other with witty put-downs, topping each other with ever more outrageous tall tales, or monopolizing the floor with virtuosic storytelling performances (Bauman 1972). The session that the Old Country Fiddler recounts in this recording is notably true to form, as participants frame their turns in assertive, critical, and oppositional terms vis-à-vis antecedent ones. The storekeeper, Sile Haskins, doesn't simply make his point; he introduces it with an authoritative "Of course"; Ed Jones retorts with "any fool could see with 'is own eyes"; Jock Sullivan backs up his claim to his son's problem-solving ability with an offer to bet on it; Jim Sanders offers another "of course"; Garl Maston puts all of them down by insisting that "warn't none of 'em got it right." These men all come on as pretty sure of themselves, pushing their own positions and putting down everyone else's in what amount to communicatively constituted displays of assertive masculinity.

While this assertive and contentious dynamic matches what we know of the keying of talk at the general store, the oppositional unfolding of the narrated event on this particular evening gains impetus from its interdiscursive link to the antecedent debate at the lyceum, apparently rather drawn out and heated. The lyceum marks another prominent site in the cartography of country communication, with distinctly New England resonances. Founded in the late 1820s for the educational and moral cultivation of the citizenry through programs of lectures and debates, the lyceum played an important role in the public culture of communities throughout the country, especially in New England and the Midwest. "In lyceum debating," as Angela Ray (2005, 28) describes it, "controversial topics were introduced and articulated, albeit in a highly formalized, agonistic way. Each question had two and only two sides; at the end of the evening, debate ceased when a decision was imposed through adjudication." For the most part, debates began to wane in importance by the 1840s (Ray 2005, 47). While some lyceums continued to stage debates through the latter part of the nineteenth century, lyceum programs by the early twentieth century consisted largely of commercial entertainments for spectatorical enjoyment, no longer locally generated and participatory but organized by bureaus that specialized in furnishing platform performers for the purpose, like Taggart himself (Ray 2005, 5, 33–34, 40–41). Certainly by 1916, when Taggart made "The Old Country Fiddler on Astronomy," lyceum debates were largely an anachronism, still recognizable in popular memory as a feature of rural community life in the relatively recent past but no longer an active feature of the American public landscape. This lag suggests an important feature of country communicability, its characterizing temporality: it is behind the times, out-of-date, still operating in ways that the audience members have presumably left behind but still recognize and remember.

From the early days of the institution, modern science was a central focus of the lyceum educational mission (Ray 2005, 4), so astronomy was a perfectly understandable theme of the Pineville lyceum debate that stimulates the further conversation at Sile Haskins's store, which serves to domesticate the more esoteric subject matter of the lyceum debate. The framing of the debate by a resolution that sets up two opposing positions is also consistent with the conventions of lyceum debate. The terms of the debate, though, are ridiculous, not based on serious astronomy but on a scientifically pointless counterposition of the relative virtues of the sun and the moon. Pineville is light years behind the leading edge of modern astronomy.

Sile Haskins's opening statement comes closest perhaps to the motivating spirit of the lyceum in demonstrating in homely, down-to-earth terms how the solar system works. This is science brought to the level of the ordinary citizen, a manifestation—albeit in terms of groceries—of the circulation of knowledge to laypeople outside the mainstream. Garl Maston has it pretty much right as well, if in a simplified recollection of an astronomy book that has come his way, as book learning makes its way into the country. The others, though, talk foolishness, ignorant of Copernican astronomy. Jim Sanders's argument that "of course the moon did more good'n the sun, for the moon shone'n the night, when it was dark, and the sun shone right'n the daytime, when y'didn't need it," is stereotpyical bumpkin humor, a noodle-tale non sequitur playing on the foolishness of country people. Ultimately, though, the conversation leads to a kind of vernacular cosmology, ruminations on the vastness of the universe and the insignificance of Pineville in it all, again not incompatible with the lyceum charter to stimulate the intellectual and moral life of communities. Clearly, there is an intellectual life in this country community, and a lively intellectual discourse, but the principal dialogue is between watered-down, vernacularized science on the one hand and credulous folk belief on the other.

Interestingly, the Fiddler closes with—and endorses—a passage from *The Course of Time*, by the Scottish poet, Robert Pollok (1798–1827), which captures the essence of Enlightenment-era, progress-oriented conceptions of customary communication in which tradition-bound participants have only a limited stock of ideas but tell them over and over again in conventional contexts throughout their lives, neither learning nor expressing anything new "from youth till hoary age." And, as if to exemplify the Pollok Principle, "There's a good'eal t'everything," as old Sylvester Boynton used to say, brings closure to the account.

CONSTRUCTING THE CHRONOTOPE OF DIFFERENCE

Clearly, the image of the communicable world that emerges from Taggart's recorded performances is not one of an isolated, self-contained rural domain. The communicable cartography of Pineville and its environs, in which he grounds his performances, is always portrayed in dialogue with an outside world, from which come lyceum performances, summer visitors, and women's suffrage lecturers, as well as books and other media that come into Pineville from their distant sources of production. And as we shall see later in this chapter, the permeable boundaries between country and city were also strongly marked for Pineville folks by the out-migration of the community's young people to the city, exemplifying the great population shift from rural communities to urban centers that was a key demographic feature of early twentieth-century American life.[23] Taggart's performances commonly revolve around a dialogic tension between the community-based, rural, vernacular modes of expression and the complex, urban, modern discourses found in books, learned in school, or experienced in encounters with strangers. That is to say, in Taggart's popular performances, as in most "rube records," country life is constructed dialogically, in contrast with the modern forms and practices that render rural life the focus of nostalgia or ridicule. The chronotope of rube humor always has a dual aspect, built by contrast (Gal and Irvine 2019, 153, 229).

"The Old Country Fiddler and the Book Agent," for example, opens with one of the most widely used and familiar tropes in the American expressive repertoire—recall "The Arkansas Traveler"—the encounter between the country bumpkin (or so he appears), fiddling away on his front porch, and the city slicker, out of his element, who appears at his door. Our particular visitor, in this routine, is a book salesman, "with a four-pound volume o' literature under 'is arm" and an altogether unprepossessing appearance, notwithstanding his store-bought urban outfit. Here is the opening episode of the Old Country Fiddler's account:

> **Charles Ross Taggart, "Old Country Fiddler and the Book Agent"**
> **(Victor B-16725-2; https://purl.dlib.indiana.edu/iudl/media**
> **/q47r66qz89), Oct. 28, 1915**
>
> One day last Spring, I was a'settin' on my porch, a'fiddlin away, when I see a feller that I took to be a book agent comin' up the path, with a four-pound volume o' literature under 'is arm. He was a kind of a spindlin', sickly lookin' chap, with a green hat pulled clear down t'the tops of his ears and a suit of store clothes on. He didn't look as if he had gumption enough t'sell peanuts to a boy. But he stepped up as pert as a rooster and said, uh, "Good morning, Ruben! I've got a book here on raisin' calves I thought you'd like to look at."
>
> "Well," I says, "if you got a book on raisin' calves, you better make a present of it to your parents. I think t'would be prof'table readin' for 'em!" Heh heh. But land sakes, 't'warn't no use. He couldn't see the point.

The book peddler was an important cultural agent in American frontier and rural life, a stranger, indexing an urban center of book production, who appeared unpredictably in the community and opened up to relatively isolated rural folks a small window on the outside world (Zboray 1993, 37–38). By the middle of the nineteenth century, with the expansion and improvement of distribution networks in the United States, the role and importance of the traveling bookseller had declined significantly, making our book agent an index of a largely bygone period, a relic of an earlier time.

But here he is, true to type—"pert as a rooster" and full of brash self-assurance—greeting the Fiddler as "Ruben," the quintessential stock appellation (together with its abbreviated nickname form, "Rube") for the bumpkin in American popular culture, and offering him a book on raising calves. The Fiddler is affronted, whether by the patronizing term of address or the young city slicker's presumption in trying to instruct him—an older rural man—in animal husbandry, or both. His witty rejoinder, "If you got a book on raisin' calves, you better make a present of it to your parents. I think t'would be prof'table readin' for 'em," goes right over the bookman's head: "'T'warn't no use. He couldn't see the point." The bookman comes off second best, though he doesn't have the sense to realize it. Rural wit and indirection encounter a conspicuous lack of uptake on the part of the cocky stranger. Here, then, is a contrast between urban and country communicative styles made manifest: urban disrespect and obtuseness versus country wit and indirection. Moreover, witty rejoinders like the Fiddler's response to the book agent play a further role in the communicative economy of the community. They are reportable, good resources for recasting as anecdotes. They make a good story, as witness the very performance before us. And, as Bakhtin (1981, 164) suggests, "the device of 'not understanding' is a useful means for making manifest a clash of chronotopes."

Uninterested in the book, the Fiddler sends the young book agent on his way. When the bookman calls on the Widow Southby, where the Fiddler's wife is visiting, he encounters another characteristic feature of rural communication: the effort to assimilate one's interlocutors to the network of locally known individuals, a work of social contextualization that is discursively accomplished: "When the Widder found 'is name was Bunker, she asked him if he was any relation to old cross-eyed Pete Bunker, used t'live over on Goose Meadow." The Widow Southby engages here in the quintessentially rural strategy of linking the stranger at her door to the familiar people and places of her own community. The book agent's surname, Bunker, provides a potential link by kinship to a familiar individual, identifiable not only by name but also by generation (old), residence (Goose Meadow), and by a distinctive feature of personal appearance (cross-eyed). The Bunker link opens up a broader connection to "the whole Bunker family" and its history, consisting, presumably, of narratives about various family members, their genealogies and alliances. The Widow's talk, then, is locally contextualizing. It is deeply grounded in the social networks, temporalities, and cartographies of Pineville and its environs. She is one of those senior women, recognizable to all rural and small-town people, who serve as custodians and recounters of community traditions. Her female effort at social connectedness stands in marked contrast to the assertive individuality of the Pineville men at the general store.

If the Widow's communicative efforts are interactional, locally grounded, and highly contextualized, what the book agent has to offer is decidedly not. While the book agent first offers his book to the Old Country Fiddler as a treatise on raising calves, we soon learn that its full title is *Carter's Complete and Comprehensive Compendium of Indispensable Information*. The book is comprehensive indeed:

> Besides the treatise on raisin' calves, it told you what t'do when y'lose yer
> jackknife, an' how many men, takin' hold o' hands, t'would take to reach
> around the Atlantic Ocean, what t'do fust after y'get drownded, 'n' how
> high all the mountains in the world'd be piled up on top o' one another, and,
> uh, what t'do when the whiffletree breaks, an' how t'run a sewin' machine,
> how many Democrats there are in Mississippi, 'n' how many Republicans in
> Vermont, an' most anything anybody wanted t'know.

It turns out, though, that the Fiddler does not in fact want to know all those things. "Well," he says, "I didn't seem t'hanker fer all that information, jest that minute." Even his partial catalog of the book's contents reveals that it is a mass of conspicuously decontextualized information, a great hodge-podge of unrelated stuff, even more so for being contained in the text object represented by the book, detached from the kind of unmediated interaction out of which the Widow Southby's grounded information emerges.

Ultimately, the book peddler's sales efforts in Pineville are thwarted at nearly every turn: "So the fella got mad and went off." The Old Country Fiddler, musing on the problem, sums up the situation in a wonderfully pithy and felicitous observation: "I reckon the Pineville folks didn't relish quite s'much promiscuous information all 'n one dose. Fact is, I didn't m'self." The promiscuity of the book's information is precisely the problem: resistant to attachment, random, indiscriminate, or, as the *Oxford English Dictionary* would have it, "Consisting of assorted parts or elements grouped or massed together without order, mixed and disorderly in composition or character." One could hardly think of a better term to capture the contrast between the literate, urban-based, modern communicative ideology represented by *Carter's Complete and Comprehensive Compendium of Indispensable Information* and the interactional, rural-based country communication represented by the Widow Southby's efforts to capture the book agent in Pineville's contextual web.

PINEVILLE PUBLIC LIFE: THE OLD COUNTRY
FIDDLER ON THE SCHOOL BOARD

A special case featuring city folks in the country that illuminates an important dimension of Pineville communicability is played out in "Old Country Fiddler on the School Board." School board meetings represent a significant site in the communicable cartography of Pineville. Here, selected members of the community—once again all male—acting in their official capacity as board members, convene to govern one of its most important public institutions. Taggart made two recordings representing school board meetings, the first of which adds most richly to our exploration of country communicability. The recording recounts how the Pineville School Board contends with a problem of discipline in the local school.

Charles Ross Taggart, "Old Country Fiddler on the School Board" (Victor B16722-1; https://purl.dlib.indiana.edu/iudl/media/494v53rx71), Oct. 28, 1915

I'm gettin' pretty well mixed up with the educational institutions of Pineville. Last March meetin' they put me on the school board, and I hadn't been in service more'n two weeks when we had that rumpus with Bill Calkins's boy.

Y'see Bill Calkins was a Pineville boy, 'n' he went off t'New York a dozen years ago, an' started a brokers' factory on Wall Street, where they make stocks and bonds, whatever they be, an' his wife has been comin' t'Pineville summers every since.

Well, this year she brought'er ten-year-old son, Archibald, and he went to the summer term o'school.

Well, the schoolmarm couldn't make young Archibald mind worth a cent. Y'see, he'd always had a private tutor up t'this time, and his ma said the tutor tried once to make 'im mind, but 't'warn't no use. Oh, land, no.

Y'see Bill's wife had a lotta modern notions 'bout bringin' up young ones. She went over t'the school one day and told the schoolmarm not to try to make 'im mind, 'cuz he warn't used to it. She was afraid 't'would jar his sensitive nature. She said that Archibald was like a flower that was just openin' out into this cooold world, and that his little subconscious self must be developed by lovely environments, and that he must be led, and not driv.

Well, the schoolmarm said she'd tried t'lead 'im, but he hung right onto the seat! Heh heh! Y'see, Archibald was of a mind t'set where he was a'mindta. And the schoolmarm, she wanted'im to set where she wanted'im to.

Well, the upshot was that a meetin' of the board was called, t'see what would be did. I expressed my sentiments good 'n' plain. Oh, I tell ye.

Oh, I mighta been a leetle mite prej-u-diced against this new system o' trainin', but that Calkins boy looked t'me like he'd had three meals a day for ten years, and enjoyed 'em, and he didn't remind me none o' no delicate flower, nuther.

I told 'em at the board meetin' it was this very spirit in so-called up-to-date sassiety that was breedin' anarchy in these here United States of Ameriky, and if the boy didn't mind the law of his home and the law of his school when he was young, he wouldn't mind the law of the state and the law of the nation when he growed up. Oh, I give 'em some oratory, now, I tell ye.

Land sakes, I knew old Judge Calkins when 'e used to run a tannery, down on Mink Meadow, and I knew that if he was alive, he'd go over t'school and tell his obstreperous grandson t'mind the schoolmarm in every particular, an' he would give her authority t'tan his hide good 'n' thorough if he didn't.

Well, I made a motion t'this effect: that young Archibald Calkins should set where the schoolmarm told'im t'set. And that if he didn't jump 'n' mind immediate, that the school board would authorize said schoolmarm to take hold of said boy, not mentally an' spiritually by his subconscious self but materially an' physically by the coat collar, and assist in the transfer.

The motion was carried and we ain't had no further cases of discipline in the Pineville Village School. [Plays a fiddle tune: "The Girl I Left Behind Me"]

The crux of the problem, the practical matter that then defines the agenda of the meeting, is the behavior of young Archibald Calkins, the ten-year-old son of a Pineville man who has moved away to become a stockbroker in New York and is thus emblematic of the great out-migration of young people from rural areas. As David Danbom (1979, 19) notes in his important study of the industrialization of American agriculture, "Rural people seemed to expect the brightest young people to make their successes in the city." But now Archibald has come to Pineville with his mother to spend the summer in his father's hometown. That is to say, in the generational transition from father to son, the Calkins family has been transformed from native Vermonters to "summer visitors." This is a telling distinction. It was during this period that Vermont was strongly engaged in developing its nascent tourist industry in a tacit counterstatement to narratives of rural decline, cultivating and marketing an image of rural New England as a reservoir of the "traditional values" that were sacrificed to urban progress and modernity (Brown 1995; B. Harrison 2005). The influx of summer visitors, then, had the potential to bring this symbolic opposition into play as they encountered the native residents and institutions of Pineville. In this instance, the problem comes to a head as young Archibald is to attend the summer term of the Pineville School. His mother, "who had a lotta modern notions 'bout bringin' up young ones . . . went over t'the school one day and told the schoolmarm not to try to make 'im mind, 'cuz he warn't used to it."

Mrs. Calkins's modern notions about discipline are a caricature of the ideas advanced by progressive educators of the day who held that children must be allowed to flower on their own but could only do so in an environment that freed their sensitive natures, as the Old Country Fiddler put it, from stress and anxiety. The "modern notions 'bout bringin' up young ones" that Mrs. Calkins attempts to impose on the Pineville schoolmarm index a much broader intrusion by "up-to-date sassiety" into rural education and country life more generally. During the second decade of the twentieth century, in precisely the period when Taggart made "Old Country Fiddler on the School Board," there was a widespread sense on the part of political leaders, journalists, educators, and progressive reformers that the great agrarian foundation of the nation was in crisis, vitiated by depopulation as young people left the country in droves for the city and enervated by conservative resistance to modern rationalized, mechanized, and capitalized modes of production, distribution, consumption, and communication (Bailey 1911; Bowers 1974; Chambers 2000; Conforti 2001; Diner 1998; Hahn and Prude 1985).

Perhaps the most prominent response to this growing sense of crisis was President Theodore Roosevelt's establishment, in 1908, of a commission to address what he deemed "the problem of country life" (Bailey [1911] 1917, 43).[24] Roosevelt's charge to the commission, founded on his conviction that "the social and economic institutions of the open country are not keeping pace with the development of the nation as a whole" (41), focused special attention on educational institutions: "You will doubtless . . . find it necessary," he wrote, "to suggest means for bringing about the redirection or better adaptation of rural schools to the training of children for life on the farm" (46). Or, as summed up in Taggart's second school board recording, the Pineville School should "spend the time learnin' our boys howta raise corn an' the gals howta cook," a program that became, ultimately, the mission of the 4-H and Future Farmers of America. Not surprisingly, the commission endorsed Roosevelt's emphasis on education, asserting in its report that "it is recognized that all difficulties resolve themselves in the end into a question of education" (121).

In the efforts to identify the root causes of the crisis, one element that came to the fore in the writings of educational reformers was the rural school board. Ellwood P. Cubberly, among the most active and influential school reformers of the day and a vigorous proponent of the rationalization, professionalization, and centralization of school administration, maintained that "one of the most serious obstacles to educational progress in the rural schools is presented by these hundreds of school trustees, who, as a rule, know little about educational needs or progress. As a body they are exceedingly conservative, and hard to educate" (Cubberly 1912, 7; see also Foght 1910, 36). This, then, was the public context within which contemporary audiences would have heard "Old Country Fiddler on the School Board."

The specific "rumpus" for which "a meetin' of the board was called" was triggered by young Archibald's refusal to sit where the teacher wanted him to. The Old Country Fiddler's account of the meeting itself consists largely of a report of his own contributions to the proceedings. He reports his "oratory" and the follow-up motion to the board in indirect discourse, though both accounts carry along with them certain characteristic features of the respective genres as performed in their original context, making the report, in effect, a performance within a performance. The oratory speaks to matters concerning the civil life of the community, the state, and the nation; espouses a strong ideological position; expresses heated affect; and builds to a formal climax, a peroration, poetically framed in parallel lines, all recognized and conventional features of the oratorical performances of the day.

The rhetorical thrust of the Fiddler's oratory aims at the core tension between the traditional culture of the rural community and the lifeways of "so-called up-to-date sassiety," reinscribing and underscoring that foundational principle of modernist social ideology (see, e.g., Simmel [2009, 601–620], Tönnies [2001]). In "Old Country Fiddler on the School Board," the tension is resolved in favor of the "old-fashioned method" of discipline and civic values, effected by the second of the Fiddler's speech acts, the motion he presents to the board.[25] He builds to the motion by recalling "old Judge Calkins," Bill's father and Archibald's grandfather, the personification of Pineville morality as an elder and as a judge. The generational trajectory from the judge, pillar of traditional authority, to Bill, who left Pineville to become a stockbroker in New York, to Archibald, paragon of modern permissiveness, summarizes the epochal shift from traditional rural life in Pineville to the morally compromised ways of the big city.

The Fiddler's motion, marked again by parallel structures, built on a more formal register of abstract epistemological oppositions and bearing one of the most recognizable hallmarks of the genre in the legalese of "said" as an adjective, amounts to a reassertion of the judge's traditional mode of child-rearing. Exercising the board's power to "authorize" the teacher to use physical force in moving the recalcitrant pupil "to set where she wanted 'im to," the motion addresses the problem of discipline in the Pineville School by adopting a policy directed at a single recalcitrant student. This is not the most efficient mode of school governance by the standards of the reformers, who would have advocated the formulation of a general institutional policy on discipline rather than a one-time fix (Betts and Hall 1914, 16; Foght 1910, 36). Still, by ratifying the motion, the board affirms the sway of traditional discipline in its domain: as long as he is in Pineville, Archibald will be compelled to submit to traditional authority. Moreover, the closing lines of the recording imply that the stern correction of young Archibald has served as an effective deterrent to other potential miscreants in the school.

In this instance, the conservative people of Pineville feel that they have made a successful stand against the incursion of modern notions from outside. Still, notwithstanding the solution devised by the board to control a single child on this one occasion, the condition on which the performance turns—the new ideologies and ways of life taken up by our children when the move away from the country and from their parents—will not go away.

COUNTRY FOLKS IN THE CITY

Urban visitors, like our hapless book agent or Mrs. Calkins, bring city and country into dialogue by showing up in Pineville and engaging with the local folks. In these encounters, the people of Pineville are on their own ground while the stranger is out of his or her element and just doesn't get it. But the visits of city folks to the country have a reciprocal pole. What happens when country folks go to the city? And go to the city they do. The first decades of the twentieth century, the years of Taggart's active performance career, marked a period of burgeoning rural-to-urban migration in the US. The impact of this demographic process was felt especially keenly in rural New England, as growing numbers of young people left the region's declining farms and sleepy towns for the educational and economic opportunities of the city. "It seems as though they can't be contented in the country any more," laments the Old Country Fiddler, "but they got to go a'traipsin' off t'the city," imbuing the city with a chronotopic cast toward youthfulness. One of those to make the transition from Pineville to the city, like Bill Calkins, was the Fiddler's own son, John, who "went t'work" in a "big store"—not a general store—in New York. His father "went down t'see 'im," and that visit provided the core of several of Taggart's recorded narratives that represent country communication in its encounter with city ways of speaking.

One revealing encounter occurs at the hotel where the Old Country Fiddler seeks lodging on the night of his arrival in the city. Inquiring about accommodation, the Fiddler finds himself in a series of puzzling exchanges with the desk clerk:

From Charles Ross Taggart, "The Old Country Fiddler in New York"
(Victor B-15532-2; https://purl.dlib.indiana.edu/iudl/media/d17c780b38),
Dec. 21, 1914

I went up t'that clerk an' I asked 'im if I c'd get a place t'stay overnight. An' he says, "European or American?"

I says, "American. Born 'n' raised here. Never 'n Europe in m'life."

Well, he says, "Uh, you want a bath?"

Well, I says, "That ain' none o' your business, mister, whether I do or not. I swan. I took a bath 'n' changed m'clothes just before I left home, but t'warn't any o' his business."

Well, he says, uh, "Outside 'r inside?"

Well, I says, "I usually make a practice takin' a bath on the outside."

He says, "I mean d'you want an inside or an outside room?"

"Oh," I says, "well, that's differnt. I'll take an inside if it's all t'same t'you. Looks a little mite stormy on the outside."

Well, he says, "Inside'll cost ya a dollar, 'n' outside a dollar'n a half."

Well, I says, "Mister, this's the fust place I ever struck where it's cheaper t'stay in th'house than 'tis outdoors!"

This time, clearly, it is the country man who just doesn't get it. The humorous dynamic of the Fiddler's encounter with the hotel clerk stems from his unfamiliarity with the features of modern, urban hotels and the lexico-semantic field that surrounds them, again an instance of "not understanding" in the service of giving substance to a clash of chronotopes. These matters are so routinized for the clerk that he refers to them in shorthand. Instead of asking the prospective guest if he wants to book a room on the American plan (with meals) or the European plan (without meals), he simply says, "European or American?" "Do you want a room with a bathroom?" becomes "Do you want a bath?" The choice of a street-facing versus airshaft- or alley-facing room is reduced to "Outside 'r inside?" Even when explained—"I mean d'you want an inside or an outside room?"—the query remains ambiguous. Does he mean inside the hotel or outside the hotel? Only someone familiar with the ways of urban hotels would understand the clerk's abbreviated queries, and the Fiddler does not. He takes the clerk for nosy or foolish, when it is he who doesn't get it.

This dynamic is part of a general pattern in country communication when it comes to the city: the country person's ignorance of characteristic features of modern, urban life and their attendant ways of speaking leads to misunderstanding, misconstrual, and talking at cross-purposes, whether it is checking into a hotel, dealing with a bellhop, eating at a restaurant, dealing with a baggage porter, and so on. The morning after his arrival in New York, for example, as enacted in another recorded routine, the Old Country Fiddler phones his son from the hotel at which he has spent the night.

Charles Ross Taggart, "Old Country Fiddler at the Telephone" (Victor 18003-1; https://purl.dlib.indiana.edu/iudl/media/049g15j212), June 21, 1916

Hellooo, Central? Say, I want to talk with my son.

Eh? My son.

Yes, he's here in New York City, been here a year and a half.

Number? Oh, just one, all the rest are gals.

What? His number? Whose?

My son's?

Heavens to Betsy, he ain't in jail, is he?

What's that?

Information? What about?

Oh, his name is, uh, John Jackson. 1500 West 86th Street.

What's that? H. O. J. Hollins?

Who's that? Oh, that's prob'ly the man he works for.

Hello. Hello! Who's this?

Oh, Central, got back to you, have we? Well, say, Central, I want to talk with Mr. H. O. J. Hollins.

Hello, is this Mr. Hollins?

Oh, this is you, is it, John?

Hello, John. Say, this is me, Dad.

In the rural and small-town telephone exchanges of the period, the operators knew everyone on the circuit and their networks of relationships. Such local familiarity allowed users simply to identify the party to whom they wanted to speak—"I want to speak to my son"—and the operator could make the connection. In the city, however, the assumption of familiarity did not hold. Parties are identified by telephone number, not kinship or given name or any other gemeinschaftliche identity feature. Like the hotel clerk's "European or American?" query or "Outside 'r inside?,'" the telephone operator's shorthand "Number?" (which we infer from the Fiddler's repetition of it [see Bauman 2010]) is twice misconstrued by the Old Country Fiddler: first as a query about the number of sons he has and then as a number assigned to jail inmates. His lack of comprehension is summed up economically in his plaintive response to the operator's further attempt at clarification: "Information? What information?" Shades of the Old Country Fiddler's encounter with the book agent. Once again, it is a matter of orders of information and the disjunction between country and city epistemologies. The Fiddler brings to his telephone conversation a personalized, community- and kinship-based orientation, only to encounter the operator's insistence on identifying his son, John, by an impersonal number. This is not necessarily the promiscuous information that the book agent is peddling, but it is fully as decontextualized, honed down to the fine point of a telephone connection, with all the indexical resonances of Pineville stripped away.

Of course, the Old Country Fiddler's engagement with the operator is also an engagement with the new communicative technology of the telephone. The operator is, in effect, a part of that technology, a component of the communicative means that the telephone represents and one of the determinants of its communicative affordances. The Old Country Fiddler's inability to deal competently with her marks him also as incompetent in the management of modern communicative technologies, at least in their most modern—which is to say, urban—guise.

NEW COMMUNICATIVE TECHNOLOGIES

Interestingly, Taggart's recorded representations of the Old Country Fiddler's difficulties with new communicative technologies also include struggles with the phonograph, the very technology on which his performance depends. In "Uncle Zed Buys a Graphophone," the Fiddler (here called by his name, Uncle Zed) and his wife first determine to buy a record player in order "t' be right in style." Even up in Pineville, that is, they have been caught up in the nascent consumer culture coalescing, in the early years of the last century, around the phonograph as the first industrially produced form of home entertainment and as an item of conspicuous consumption, a commodity of distinction (Kenney 1999, 28, 52–56; Millard 1995, 35–64).

Charles Ross Taggart, "Uncle Zed Buys a Graphophone" (Columbia 78206-1; https://purl.dlib.indiana.edu/iudl/media/x91366cx8x), Dec. 1918

[A couple of bars of fiddle tuning check]

Say. Marm and me decided that if we wanted t'be right in style, we better get one o' these here hail Columbia happy land talkin' machines. So I hitched up the old mare and went over t' New'bry Street 'n' bought one.

I'd just got it loaded into the hind end o' m'wagon, when that clerk followed after me 'n' says, eh, "Hey, there, mister! You need some records t' go with that machine."

I says, "Hold y'hosses, son. No extras f'me this trip. I may get some o' them later. Git up there, Jeremiah!" An' I drove off 'n' left 'im.

Well, that evenin' Marm an' me thought we'd have a little music. So I wound the thing up and sot it a'whirlin'. But it didn't make no sound! Not a sound. I put my ear down and listened all over it, but I couldn't hear a thing. Not a thing.

Well I was mad. An' I went t'the telephone 'n' called up that clerk, 'n' I says to him, "Say, young fella, that there hail Columbia happy land talkin' singin' machine o' yourn don't make no noise! How in Sam Hill do you make it go?"

Well, he sassed me right t'my face, 'n' said it I hadn't a'been in such a darned hurry to go, he was a'goin' t'tell me. He said I had t'have some se-lections 'n' some needles t'make it work.

"Needles?" I says. "Young man, you've made a leetle mistake. I bought a singing machine, an' not a sewin' machine."

Well, he said it wouldn't sing, nor play, nor talk, nor nothin' without needles. So I told 'im t'send me a good lively tune an' a needle an' a spool o' thread t'sew it on with an' I'd try it again.

Well, when we finally got it t'goin', it played a mighty good tune. I swan t'man if it didn't swing along somep'm like an old camp meetin' tune we useta sing. The name of it was, uh, "Hesitation Waltz."

Well, we played it 'n' played it, for about a coupl'a weeks, 'n' got it most learned b'heart.

An' one day George Macklin come in 'n' says, uh, "Uncle Zed, why doncha play the other side o' your tune?"

Well, I never thoughta that, 'n' he flopped that over on its back 'n' played another tune, most good's the fust one. That was called "One Step." But by

Jiminy Crickets, it sounded t'me like a whole flight o' stairs! Hah, hah, hah!

Well, one day I was down to Bradford with a load o' hogs, an' I was a'gittin' my hair cut in the barber shop, when all of a sudden, I heared my "Hesitation Waltz" a' goin'.

Well I jumped up with that apron 'round my neck an' shouted, "Somebody stole my tune!"
But the barber says, "Oh, siddown! Siddown. I guess yer tune'll keep till y'get y'hair cut, won't it?"

Well, after I got all trimmed up, I listened all 'round, but it had stopped. An' I couldn't find it nowhere. But the most curious thing about it was, when I got home, there was my tune, right on the machine where I left it. Talk about yer mysteries!

Well, Mariah says prob'ly they made two needles just alike, by mistake, but I believe t'was the spirits.

Wanna hear a tune?

[Plays "Hesitation Waltz"]

As Taggart was a Columbia Record Company performer in 1918, at the time he made "Uncle Zed Buys a Graphophone," it is not surprising that Uncle Zed and his wife determine to buy a Columbia graphophone, as Columbia called its record players to distinguish them from the Edison phonograph, the Victor Victrola, and other players sold by smaller makers. As if to underscore their location way out on the margins of modern life, though, they misconstrue the brand. Picking up on the resonance between the Columbia of the brand name and the opening lyrics of "Hail Columbia," at that time the unofficial national anthem (this was well before "The Star-Spangled Banner" was adopted as the American national anthem in 1931), what they want is "one o' these here hail Columbia happy land talkin' machines" ("talking machine" was a widely used generic term). Moreover, Uncle Zed goes to pick up his stylish new graphophone in his horse-drawn wagon, about as unstylish and old-fashioned a conveyance as you can get.

And then when they get their new purchase home, it turns out that they do not know how to make it work. Indeed, in their ignorance, they undermine an essential part of Columbia's strategy, held in common by all the manufacturers of record players, of getting their playback machines into people's homes only partially as status symbols in their own right but also, crucially, as a basis for making larger profits on the sale of records and other accoutrements like needles on a continuous basis. The machines represented a one-time purchase, but the records and supplies continued to produce profits as new ones were issued and consumers were eager to keep up with new hits (Millard 1995, 44, 48). Uncle Zed, however, does not realize that he needs "se-lections" to play on his new machine, rejecting them as "extras," though they are essential to this medium of home entertainment. When he tries to get his new graphophone to play, "it didn't make no sound. Not a sound!" And this makes him mad, as we all are when some new piece of technology resists our best efforts and refuses to do what it is supposed to. His disgruntled call to the clerk, to complain about the recalcitrant machine, provokes the clerk to an equally testy response, affronting Uncle Zed still further. As far as the clerk is concerned, Uncle Zed has brought the problem on himself by rushing off with his new graphophone without heeding the clerk's insistence that he needed some records to go with it. But now he still wants to find fault with the clerk, when the latter informs him that he needs needles as well. "'Needles?' I says. 'Young man, you've made a leetle mistake. I bought a singing machine, an' not a sewin' machine." And even after it all gets sorted out and Uncle Zed gets the machine to play his new record, "Hesitation Waltz," he doesn't realize until a younger and more up-to-date neighbor, George Macklin, suggests it that his record has two sides and he can play a second tune if "he flopped that over on its back."

What we see in these exchanges between the Old Country Fiddler and his younger interlocutors, George Macklin and the salesclerk, are the contours of a very powerful trope in the construction and rhetoric of consumer culture, especially in relation to communication technologies and home entertainment: they are oriented toward youth, while the old folks don't get it and just get cranky when they try to get the damn things to work. (I write this out of ample personal experience.) Uncle Zed, of course, is doubly handicapped: he's not only old, but he's country to boot. His efforts to be "right in style" are doomed from the start.

It is worth noting in this connection that the Old Country Fiddler's ignorance and ineptitude concerning new technologies are of more general scope. In one of the recorded routines in which he recounts his experience in a New York hotel, discussed a bit earlier, he has trouble with a couple of other appliances (besides the telephone) that he encounters in this urban environment. First, he tinkers repeatedly with a curious little button on the wall of his room, not realizing that it's the buzzer for summoning the bellhop, who keeps showing up unexpectedly in answer to his summons. Then, when he's ready for bed, he can't seem to extinguish the electric light in his room—it won't go out no matter how hard he blows on it—and winds up lowering it on its cord so that he can shut it away in a bureau drawer. This was a widely repeated stock routine (Marvin 1988, 219), used to poke fun at country people in urban environments, bewildered by modern technologies.

Nor is it only his failure to understand the mechanics of the graphophone—that is, how to get it to play—that marks the Old Country Fiddler's ignorance concerning the new medium. Perhaps more interesting is his struggle to comprehend the affordances of sound recording as a technology of cultural reproduction. First, in ordering the record he needs to make his graphophone work as it should, he foregrounds not the material object but the music it will produce, asking the clerk to send him "a good, lively *tune*." Then, as a musician, he plays the record over and over until he has learned the tune himself. The recording thus becomes a source of a new piece that he can add to his own performance repertoire; his ownership of the recording thus amounts to ownership of the tune and the right to play it himself. To be sure, playing the same record intensively, again and again and again, also serves to undermine the extensive consumerist logic promoted by the record companies, who wanted customers to buy many records, as they were issued (Gitelman 2006, 63–68).

Most revealing, however, is his reaction when, sitting in a barber's chair in Bradford, where he has gone to sell some hogs, he hears the very tune he has purchased coming from somewhere nearby, presumably being played on someone else's record player. Or so we assume, knowing what we know about the medium. The Old Country Fiddler's reaction is that "somebody stole my tune!"

Consider what must underlie such a notion. By his understanding, the Old Country Fiddler has bought and paid for the tune, and thus he owns it. It is not the record as a material commodity that counts; it is the tune, as a piece of music, that the recording represents to him as a piece of property and that is uniquely his to activate on his graphophone and to learn himself. For someone else to play it amounts to an act of theft. Unable to find the source of the music he has heard in Bradford, he fears it is lost to him. He is surprised and mystified, then, to find that "when I got home, there was my tune, right on the machine where I left it. Talk about yer mysteries!" The only explanation he can come up with, falling back on the age-old traditional way of explaining the otherwise unexplainable, is that hearing the tune in Bradford while it was still at home in Pineville must have been the work of the spirits.

This is a remarkable representation. What Taggart has seized on in this depiction of the Old Country Fiddler's understanding of the medium of phonography and his way of accounting for the "mystery" of hearing his tune in another town amounts to a popular representation of what Walter Benjamin termed in a more esoteric philosophical register "the work of art in the age of mechanical reproduction." The phonograph is a technology par excellence of mechanical reproduction, one of the key technologies that Benjamin himself identifies with the new age, and Taggart's recordings date from a period just following the moment that Benjamin defines as a watershed, in which technical reproduction "captured a place of its own among the artistic processes" (Benjamin 1955, 219). When Uncle Zed hears his tune in an unexpected place, far from home, he is experiencing precisely the disjunction that Benjamin sees at the core of the reproduction, its lack of "presence in time and space, its unique existence at the place where it happens to be" (Benjamin 1955, 220). The reproduction, a replication detached from the spatiotemporal specificity and indexical resonances that accrue to the unique work of art in its context, is as promiscuous as the decontextualized information gathered in the book agent's volume. Uncle Zed's experience thus marks him as the child of an era before the age of mechanical reproduction, when the piece of music, or any work of art, was anchored in its proper context, back home where he left it. His wife has at least an inkling that mechanical reproduction is to blame for the mystery, even though she attributes the replication to the existence of two needles rather than two records and considers that the making of "two needles just alike" must have been a mistake. Uncle Zed, though, doesn't believe it. He falls back on the uncanny, blaming "the spirits" rather than the technology of mechanical reproduction.

In addition to providing comedic fodder for his performance routines, Taggart's naive and unsophisticated orientation to the phonograph served well as an advertising and promotional device. A widely reproduced promotional photograph, discussed in more detail in the next chapter, portrays Taggart as the Old Country Fiddler listening to one of Taggart's own recordings on a cabinet model Victrola. Accompanying captions inform us that "the Old Country Fiddler hears his own voice" and "the Country Fiddler tickled with himself." Consider: he had already experienced the process of making the recording and so should not be surprised by what he hears, yet the expression on his face—especially his open-mouthed, toothless grin of delight—suggests that he still marvels at hearing his own voice issuing from the machine.

COUNTRY MUSIC

The Old Country Fiddler's fiddle, of course, is a communicative technology in its own right. To this point, I have focused on aspects of verbal communication: language, ways of speaking, modes of spoken interaction, speech events, technologies of the spoken word. But, as his performance persona makes explicit, there is another major semiotic register that Taggart foregrounded in his performances—namely, music. He is, after all, the Old Country *Fiddler*. And his fiddle, as he himself describes it on "Uncle Zed and His Fiddle," is perfectly consistent with the country persona he projects—a largely homemade object.[26] He spied the instrument at a neighbor's house, in conspicuously distressed condition: it was "layin' round all stove up" and "there warn't no top to it 'n' the pegs was gone." Determined to have it nevertheless, he "asked Dan what he'd take for it," and that move opens the way to a bargaining exchange. "Well," Dan said, "he kinda hated t'part with it. He said it'd been in the family for a lo::::ng time," and so on, building the broken-down instrument's value as best he can. The Fiddler "finally made a trade with 'im. I give 'im a heifer calf and a coupl'a dollars t'boot." Relying on his own skills at woodworking, "I made the top m'self, out of a hemlock slab, and whittled the pegs outa clothespins, 'n' got it so it sounds fust rate." In the end, he has an instrument that he values highly, calculating that, including his work and materials, "my fiddle cost me about seventeen dollars 'n' forty-nine cents, but I wouldn't take a twenty-dollar bill for it." That is, the Old Country Fiddler has acquired his treasured instrument by barter and handcraft, the old ways of acquiring material goods, in perfect contrast with his new record player, purchased for cash and commercially produced.

We have already encountered the Old Country Fiddler playing his prized instrument on his front porch as the feckless book agent approached with his hefty book under his arm. While the Fiddler takes pleasure in playing for himself, as most musicians do, it is the social enjoyment of music that figures most prominently in Taggart's recorded performances. At a minimum, the Fiddler is eager to play for whomever he may be addressing at the moment. "Wanna hear a tune?" he asks at the end of "Uncle Zed Buys a Graphophone" and then proceeds to play one, whether for some virtual interlocutor to whom he has been telling one of his stories or us, the listening audience.

Fiddle tunes figure in a variety of guises on Taggart's recordings. In their simplest form, the tunes serve to bracket the featured narratives, with brief passages—usually the A part plus the B part, played once through, sometimes the A part alone—played without any verbal framing before or after the verbal performances or both. On other recordings, the Fiddler prefaces the tunes with verbal introductions, giving the name of the tune with some additional bits of contextualizing information. "I'm goin' t'play 'Ol' Boneypart's March over the Rhine,'" he tells us at the beginning of "The Old Country Fiddler in New York," later closing the record with a reflexive coda: "I'll play ya 'Pop Goes the Weasel.' I get t'talkin', tellin' stories, I forget what I'm doin'." In a few instances, the closing tune is thematically related to the story that precedes it. At the end of "Old Country Fiddler and the Bandit," for example, following a disquisition on criminality and the coddling of criminals nowadays, the Fiddler observes, "I guess I'll have to play you the 'Rogue's March'" after that and proceeds to do so.

More significant for our purposes, however, is the Fiddler's depiction of the place of music in the public life of Pineville. In "A Country Fiddler at Home," for example, the Fiddler tells us about the Pineville Orchestra, consisting of himself on fiddle, Sile Haskins on clarinet, and his son on cello ("big fiddle" in the Scottish tradition), who played "sometimes here in our settin' room, sometimes in Symphony Hall, up over Sile's grocery store."

Charles Ross Taggart, "A Country Fiddler at Home" (Edison 8468-A; https://purl.dlib.indiana.edu/iudl/media/188158j593), May 26, 1922

I wish I had the rest of the Pineville Orchestra here. We'd give you some re::al harmony. I play the fiddle, Sile Haskins plays the clarinet, and my boy did play the big fiddle, 'fore 'e went off t'the city.

We was just a'gettin' so we could play pretty well. Sile got so's he could sque::al a tune outa that clarinet in pretty good shape. He said it tickled 'is lip when 'e first blew it so he couldn't hardly hold it to 'is mouth!
Hah hah hah!
But when my boy went, he took the biggest part o' the orchestra with 'im. We hated awfully to have 'im go.

We used t'have some re:::al good times playin' together, sometimes here in our settin' room, sometimes in Symphony Hall, up over Sile's grocery store.

Oh, I do wish these young folks would stay t'home, where they belong. By jiminy, it seems as though they can't be contented in the country anymore, but they got to go a'traipsin' off t'the city.

The tone of nostalgic regret that runs through the Fiddler's recollection is striking. The Pineville Orchestra, once a centerpiece of the community's musical life, is no more. When the Fiddler expresses the wish that he had "the rest of the orchestra here," it is his son that he is missing. Like Bill Calkins, the Fiddler's son, John, left his rural home and "went t'work" in New York, taking "the biggest part o' the orchestra with 'im," and "the real good times" the group provided for its members and the community now exist only in memory.

In a similar vein, in one of the most vivid of Taggart's recorded performances, "Old Country Fiddler at the Dance," the Fiddler describes the "country dances" held in Pineville's old town hall. In the course of his account, triggered during an evening of social dancing at his son's house in New York by the shock of the "entirely new kinda dancin'" he and his wife witness there, the Fiddler cannot help but contrast the "scandalous performance" by his son's guests and "the good old dances we useta have up in Pineville." The Fiddler lists off a series of traditional dance tunes, including well-known pieces in the New England and broader American repertoire: the "Virginia Reel," the "Irish Washerwoman," "White Cockade," "Yankee Doodle," "Moneymusk." The description very quickly leads to a breakthrough into performance, as the Fiddler recalls the set of tunes used to accompany the dance—"here's the way it went"—gets caught up in his own recollections, and begins to call the figures as the caller would at the dance itself, accompanied by a musical ensemble he conjures into being for the occasion:

Charles Ross Taggart, "Old Country Fiddler at the Dance" (Victor C-18015-1; https://purl.dlib.indiana.edu/iudl/media/2773865m1s), June 23, 1916

Hey! I wish you could see one of 'em! Them old-time figgers danced out good'n lightly. I c'n see 'em now in the old town hall, all lined up ready t'start. Then Sile Haskins would rosin up the bow an' start the "Virginia Reel." How they would step it out!

I c'n see Jock Sullivan now. He allus come with his cowhide boots on an' when he sashayed up an' down the middle, it would shake the whole buildin'.

Y'see, for the Virginia Reel, they used t'play the "Irish Washerwoman," "White Cockade," an' "Yankee Doodle," an' here's the way it went [ensemble consisting of piano, clarinet, and trumpet plays "Irish Washerwoman" as the Old Country Fiddler calls the figures]:
> Head to foot couple,
> sashay for'ard and back.
> Balance your partner!
> Back t'yer places!
> [ensemble plays "White Cockade"]
> Swing t'the left!
> Back to place!
> First lady jumps in!
> Sashay t'the center!
> Go it, Jock!
> Both hands!
> [ensemble plays "Yankee Doodle"]

"I tell you, folks," he concludes, "for a right down healthy, riproarious, rollickin' good time, that didn't hurt nobody, give me a regular old-fashioned country dance."

"Old-fashioned" is the key term here. The "old-time" dances that the Fiddler describes so fondly and the fiddle tunes played at those dances are rooted in the past—in fact, they are a thing of the past. What the Fiddler is recalling with such nostalgic enthusiasm is "them good old dances we *used to have* up in Pineville." Likewise, the Pineville Orchestra is no longer, and it is the Fiddler's own son who brought about its demise when he "went off t'the city." The musical culture of the rural community has thus been eroded by the depopulation of the region. With the dances, as with the orchestra, what is left is the memory of bygone entertainments.

Moreover, when all those young people from the country relocate to the city, they acquire entirely new tastes in music and dance. We have already seen the "entirely new kinda dancin'" that his son and his guests have taken up. "That's the very latest style o' dancin', Ma," the son protests when his mother scolds him for "allowin' such doin's in his house." On another occasion, when the Fiddler and his wife are present at a dinner party hosted by their son and daughter-in-law in New York, one of the others in attendance happens to be a musician as well.[27] The guest's modern piece is a slow, lugubrious, and dissonant composition, at least judging by the Old Country Fiddler's imitation of it. Invited to take a turn himself, the Fiddler "ripped off m'coat 'n' vest 'n' rolled up m'sleeves, 'n' give 'em 'Old Zip Coon'" and a series of other traditional fiddle tunes. The response of the other dinner guests is telling: "Y'oughta see 'em laugh! Jiminy! Tickled 'em 'most to death." As in so many of his other urban encounters, the Fiddler doesn't get it, taking his listeners' ridicule as an expression of wonder and delight. This episode of dueling fiddles, then, with its contrastive musical styles, tastes, and evaluations, recapitulates the rural-urban disjunctions played out in the verbal encounters we have considered earlier. Even further, it plays out a generational shift as well, as the Fiddler's son, formerly a key member of the Pineville Orchestra, has adopted new musical tastes with higher cultural capital. The chronotope of country communicability implicit in the persona of the Old Country Fiddler is succeeded in these recorded performances by its contrastive musical and choreographic temporality and cartography: new, modern, youthful, and urban.

With the demise of the Pineville Orchestra, Taggart gives us a foreshadowing of the community's musical future in "Old Country Fiddler at the Wedding."[28] Invited by young Jenny Sullivan to play the fiddle for her wedding, the Fiddler, who "knowed'er ever since she was a baby," readily agrees. When he launches into "Marchin' through Georgia" at the wedding rehearsal, though,

Jenny makes clear that that old piece is not what she has in mind. Not knowing the "Wedding March," which is what the young bride-to-be wants, the Fiddler seizes on his Victrola "talkin' machine" (reflecting his shift from Columbia to Victor) as a means of resolving the problem and orders a recording of the piece from town. Unfortunately, through a mix-up on the part of the postmaster's boy, who delivers the wrong parcel just before the wedding is to begin, the record that the Fiddler puts on the turntable turns out to be "The Irish Washerwoman," which is "a leetle might rapid for a weddin' march" and creates pandemonium among the wedding guests. When the boy returned later with the correct record, the Fiddler gave him "a piece o' my mind, now, I tell ya." Then, breaking out of the story, the Fiddler tells us, "I've got the record right here, I'll play it for ya," and it turns out to be a brass band rendition of the "Wedding March." Not only has the phonograph become a substitute for his kind of live performance in Pineville, but it now stands in for the Old Country Fiddler on his own recording. The new technology of musical production has taken a step toward displacing live musical performance.

THE PINEVILLE BAND

As consequential as the advent of the phonograph and the demise of the Pineville Orchestra may have been in transforming the musical communicability of Pineville, they were not the only forces of change that threatened to replace the traditional musical culture of the town. In fact, as the Fiddler's son, John, moved away from home, another musician moved into town, and the newcomer—an emissary from the outside, like the bookseller—proceeded to introduce a new musical institution into the public culture of Pineville, one that aligns, as it happens, with the brass band recording of the "Wedding March" that took over the musical production at Jenny Sullivan's wedding.

Charles Ross Taggart, "The Pineville Band" (Victor B-15600-3; https://purl.dlib.indiana.edu/iudl/media/r171887j14), Apr. 12, 1915

["Irish Washerwoman," 13 sec.]

We've got a brass band started up in our town. Had a feller move in't'town, knew how to play on about everything under the sun, so he got most everybody into it of any account. Wanted I should jine. I told 'im I guessed I'd wait a while, and see how the thing, uh, panned out. He wanted I sh'd play something called a clarinet. Well, I tried the thing, but land sake, it tickled m' lips so's I couldn't keep it in m'mouth!
Heh!

Well, I'se tryin' it a little bit easy one day, upstairs, 'n' Mariah come up bringin' me some salt water. She thought I'se sick t'm'stomach! Heh, heh, heh!

Well, uh, th'band, when they fust was organized, they used t'play out on the steps o'th'Methodist meetin' house, but th . . . the community they got kinda nervous 'bout it, they thought't'd give the town a bad reputation, y'know, folks drivin' through, strangers, so they asked 'em, got up a petition, asked 'em if they wouldn't play where it t'warn't quite so conspicuous. So after that, they played in Symphony Hall, up over the sawmill. That made it some better.

Well, a while ago, the women got up a kind of baked bean sociable t'raise money t'buy a new slide trombone for the town clerk, and, uh, the band, they played a piece t'the sociable. I al's supposed t'was "Boneypart's March," they played, but Sile Haskins, he said t'was "Home Sweet Home" [laugh/snort], but the leader said it's "Marchin' through Georgia," and he ought t'know.

Well, y'see they, uh, they're gonna get up a kind of a pigeont [pageant], like, they call it, out on the common, t'raise money t'buy a new bass drum, f'the blacksmith. Uh, he, uh hit it s'hard, whangin' away on that "Marchin' through Georgia," the night the sociable, it bust a hole right through it, so they got t'get a new'un.

I understand they're gonna have a new bandstand built, right out front the post office and play every Sat'dy night while they're waiting for the mail, so I guess they'll have a pretty good time on't anyway.

Got so's they c'n march a little bit, now, so the leader, he got Jock Sullivan, the Overseer o' th'Poor, t'be the drum major. And, he give 'im a great long, gold-plated broomstick with a croquet ball on top of it, 'n' told 'im t'practice with it. Well, Jock took it home, 'n' brought it back next day 'n' said he couldn't find no mouthpiece on it. Heh! Said 'e'd blowed all over it but couldn't get no music out of it. Well, the leader told 'im t'swing it 'round, keep time, so he does that 'n' gets along better now.

I'll show ya how one th'pieces goes, th'band's been learnin'. Goes someth'n like this: [plays fiddle as rhythm instrument and sings, "Bomp bom bom, bomp bon bom," with transition that sounds like "Down Yonder"].

In organizing a local brass band, Pineville joined a popular cultural movement that reached widely into American communities throughout the late nineteenth century. Having a band became in this period a prominent touchstone of community identity, pride, and commitment to the cultivation of the citizenry, a potent claim to distinction. G. F. Patton, a prominent champion of the band movement, maintained in 1875 that "it is a fact not to be denied that the existence of a good brass band in any town or community is at once an indication of enterprise among its people, and an evidence that a certain spirit of taste and refinement pervades the masses" (quoted in Camus 2013). Driving home the point, an instrument catalog issued in 1881 proclaimed, "No town or village ... can pretend to have attained much progress in social esthetics which is not blessed with a good brass band" (quoted in Hazen and Hazen 1987, 44; on the American band movement, see also Camus 2013; Cockrell 1998; Newsom 1994; and Proper 1998).

The burgeoning of local bands in the US was fostered by a series of convergent historical factors—social, technological, and economic—including the stimulus provided by returning veterans of Civil War military bands; the increasing presence of Italian and German immigrant musicians with the skills to organize, train, and maintain musical groups; improvements in the technology of instrument manufacture; and the growth of publishing houses with the capacity to produce and distribute musical scores on a wide scale. The band movement of the late nineteenth century was thus an emergent force in the development of a national-level, commoditized popular culture industry in the US. The founding of a brass band signaled to Pineville the community's entry into a broadly national arena of public culture.

The Old Country Fiddler's account of the formation and development of the Pineville Band traces a familiar pattern. To launch the process, the formation of a band requires the presence of a trained and versatile band musician, often, as in the case of Pineville, an individual from outside the community who has had the opportunity to gain the necessary expertise to make a go of it in a new locale. The leader must enlist the participation of reputable community members—individuals "of any account"—and assign responsibility for learning how to play specific instruments to players who, more often than not, have no musical skills whatsoever. That is, unlike the earlier—now defunct—Pineville Orchestra, membership in the Pineville Band requires no prior experience or musical competence. Even the Old Country Fiddler, who has well-recognized musical ability, is assigned an instrument with which he is entirely unfamiliar, as brass bands obviously had no place for stringed instruments. Unable to play the clarinet with the degree of competence that sustains his fiddle playing, he opts to remain on the sidelines. He can imitate the band's new tunes on the fiddle, but he remains an observer, not a member.

Not surprisingly, the neophyte musicians' initial lack of musical ability is clear to all, moving the community to petition the band to move its rehearsals from visible (and audible) public space on the steps of the Methodist meeting house to a less conspicuous site. Accordingly, the band shifts its rehearsals to the Symphony Hall, formerly the venue of the Pineville Orchestra before that ensemble ceased to exist (although transferred, in this recording, from the space over the store to the corresponding space over the sawmill). Still, notwithstanding its inauspicious beginnings, the people of Pineville—most prominently, the women—remain willing to support the fledgling musical addition to the artistic culture of their town. The women organize a fundraiser for the band, in the form of a bean supper, to help buy instruments for newly minted musicians, which also furnishes the occasion for the band's public debut. It appears, though, that the band is not quite ready for a smooth opening: the blacksmith beats his bass drum so enthusiastically as to destroy it in the process (requiring yet another fundraiser), and the band's playing is still not up to the standard that would make the tunes it offers—all old, familiar standards—recognizable to the audience. They stay at it, though, progressing as they go, learning to march as they play, a common step in the developmental trajectory of local bands, even as their neophyte drum major, still with the help of the leader, learns how to swing his makeshift baton in appropriate fashion. Most importantly, the band cements its place in the public culture of Pineville, gaining a durable venue for its performances, a bandstand, and a slot in the weekly rhythm of the town, playing every Saturday evening as the townspeople await the arrival of the mail. The band thus becomes fully a part of refigured Pineville musical chronotope, with a regular position in the spatial and temporal structure of the community. Where the Pineville Orchestra was part of the town's past, its "used to," the Pineville Band is vigorously of the present and reaching toward the future as its members learn more tunes and anticipate their place on the bandstand they're "gonna have," "right out front the post office."

Notwithstanding all the energy and enthusiasm surrounding the band, however, there is more than a touch of irony implicit in Taggart's recording. When "The Pineville Band" was produced, in 1915, the national enthusiasm for local bands was severely in decline. Even as the Old Country Fiddler was reporting on Pineville's newfound excitement at having its own brass band, local bands as a form of public and popular culture were themselves giving way to nascent forms of mass-mediated entertainment, including most prominently cinema and the phonograph. With the advent of broadcast radio in 1920, the community brass bands faded from the American soundscape. Once again, then, for Taggart's audiences, Pineville reaches backward toward the past, still recent enough to be remembered clearly but saturated with an aura of bygone ways of life.

CONCLUSION

As a genre, rube records were inherently chronotopic, casting their representations of rural life as comically old-fashioned, anachronistic, behind the times, and in fundamental contrast—whether implicit or overt—with the up-to-date modernity of life in the city. For the most part, however, the depictions of life in rural America featured on those early recordings were representationally thin, suffused with long-established stereotypes of country people and shaped by inherited theatrical conventions for the comic performance of life in country communities. Charles Ross Taggart's recorded performances as the Man from Vermont and the Old Country Fiddler, however, stood markedly above the general run of rube records in their chronotopic complexity, offering a richer, more nuanced depiction of a rural community in time and space within the context of early twentieth-century American life.

To begin with, Taggart's performances were more specifically emplaced than other rube records, which were only vaguely marked as country by performers who were not themselves from rural backgrounds but assumed a rube persona, indexed especially by corny names, a few emblematic features of casual, nonstandard English, and raucous laughter. Taggart's records were set explicitly in Vermont, the symbolic construction of which as a relic area was widely disseminated and familiar to urban audiences. Moreover, Taggart himself grew up in Vermont and was rooted in the culture and language of the state. His deep attachment to the place rendered his portrayals of life in Pineville more sympathetic than the representations of rural life by other rube performers. Cal Stewart, for example, who built up a substantial body of recorded stories about Pumpkin Center even more extensive than Taggart's representations of Pineville, tended to foreground breakdown and moments of chaotic disorder in his characterizations for comic effect while Taggart offered a more detailed and nuanced picture of social life in a rural community.

At the same time that Taggart was deeply familiar with life in rural Vermont, he was no stranger to city life, pursuing formal training in performance skills in Boston and living intermittently in New York in the early years of his platform performance career (Boyce 2013, 32). His urban experiences gave him sufficient distance from his rural background to develop a reflexive, analytical perspective on the differences between the country and the city. Thus, he was well equipped to develop and exploit the chronotopic resonances of Vermont rural life in the broader environment of early twentieth-century America.

An additional factor that helps to account for the richness of Taggart's recorded oeuvre is his well-established prior career as a platform performer. To be able to sustain an evening's or a full tour's worth of stage performances around his Old Country Fiddler persona required more richly conceived representations of Pineville life than brief vignettes developed from the outset to accommodate the restrictive affordances of early recording technology. Thus, Taggart had a well-crafted and practiced repertoire of Pineville-focused performances from which he could adapt his recorded narratives rather than, say, Cal Stewart, who devised his accounts of Pumpkin Center first and foremost for the talking machine. For the most part, with the exception of Stewart's and Taggart's recordings, rube records were shaped by minstrel-show and vaudeville performance formats, which revolved around snappy exchanges consisting of a setup line or two and a witty rejoinder, punctuated by exuberant laughter. While some of these records displayed a recognition of the importance of local social relations as a contextual frame in village life, the local information exchanged was brief and relatively superficial. This selection from "The Village Gossips," by Cal Stewart and Steve Porter, offers a good example:

> **Cal Stewart and Steve Porter, "The Village Gossips" (Edison Blue Amberol 1594-2; https://purl.dlib.indiana.edu/iudl/media/x02138rf2z), Aug. 1912**
>
> Jim: Hello, Josh.
> Josh: Hello, Jim, how be ya?
> Jim: Whatcha doin' with that fishin' pole? Going fishing?
> Josh: No, I'm just goin' down t'th'crick t'give this here worm a few swimmin' lessons.
> Both: Laugh.
> Jim: Smart, ain't ye?
> Josh: Well, tol'ably so. Ye hear about Lige Willett?
> Jim: No, what's he been doin'?
> Josh: Wanted t'keep 'is chickens from gettin' cold, so he put whiskey in the chickenfeed.
> Jim: How'd it work?
> Josh: Fust rate. One ol' hen got s'drunk she forgot what she was doing an' laid ten eggsthat day.
>
> Both: Laugh.
> Jim: Oh, say, did ya hear 'bout Hank Weaver?
> Josh: Don't mean t'say he's been arrested agin?
> Jim: Nooo, he was down the grussry store an' he seen that 'ere [air] new-fangled 'lectric fan a'runnin' an' he says, "Ezry, if you don't let that 'ere squirrel outa that wheel, he'll run 'is fool head off!"
>
> Both: Laugh.

And so on, in this vein.

"The Village Gossips" is a one-off performance, and while it suggests an awareness of country communicability in its attention to local social relations as a conversational resource and the general store as a site of talk in rural communities, it does not begin to compare with Taggart's extended and nuanced attention to the temporalities and cartographies of country communicability. Taggart's recorded oeuvre constructs a world in which ways of speaking and interacting and playing music are organized in terms of contrasting temporal and spatial coordinates that encompass the rapidly changing social world of rural Vermonters in the early decades of the twentieth century. But perhaps the most unique aspect of Taggart's explorations of country communicability is his extension beyond verbal forms to the place of music in the communicable chronotope of rural life. Taggart's integration of the verbal and the musical was unique in the rube-record catalogs of the day and, notwithstanding the later popularity of traditional fiddling on records, remains his singular contribution to the symbolic construction of country culture in commercial mass entertainment.[29] It is worth remarking here that although the verbal and the musical aspects of Pineville communicability that Taggart portrays inhabit the same chronotopic environment, it is only the musical culture—the Pineville Orchestra, the country dances—that the Old Country Fiddler reports as now missing from contemporary life. That is, while Taggart constructs Pineville as old-fashioned in its ways, both verbal and musical, the ways of speaking remain part of the community's current communicable profile. It is only the old ways of making music and the dances that were a key site of music making that have been lost, now relegated to nostalgic recollection.

Integrally related to Taggart's synthetic integration of verbal and musical dimensions of country communicability is his reflexive exploration of new communicative technologies as they enter into and transform the expressive lives of rural people. Taggart cultivated the phonograph especially, both as a presentational medium for his own performances and a productive complement to his work as a platform performer, but he also exploited the talking machine as a thematic resource, offering a penetrating exploration of the new medium's transformative effects on the musical culture of rural communities and highlighting the chronotopic tension between traditional rural communicability and modern consumer culture.

Cal Stewart, in his guise as the talking machine storyteller, embodied the transitional moment between oral performance as a traditional vehicle of co-present, vernacular expression and phonographic performances as mechanically reproduced, mass-mediated forms of popular entertainment. For Charles Ross Taggart, however, the calibration of that epochal transition was more subtle. Taggart engaged the transformative effects of the new medium directly, even analytically, in his disquisitions on the social and epistemological complexities attendant on the work of musical art in an age of mechanical reproduction and in his reflexive musings on how live performance in community settings becomes mechanically mediated in those same settings and becomes at the same time a mass-mediated commodity produced for home entertainment. For Taggart, more than any rube performer, there was indeed "a good'eal t'everything."

NOTES

1. Biographical information on Taggart is drawn from Greer 1927; and Boyce 2013.

2. Taggart to Harry Harrison of Redpath, January 23, 1910; contained in the Redpath Chautauqua Bureau Records, Special Collections Department, Special Collections and Archives, University of Iowa Libraries (Iowa City), series I, box 318. Used by permission. Subsequent references to the Redpath Chautauqua collection cited as Redpath Collection, [box number].

3. Redpath Collection, box 319.

4. *Agitator* (Wellsboro, PA), July 26, 1922, 3.

5. *Mansfield (OH) News*, November 4, 1933, 10.

6. Redpath Collection, box 318.

7. Taggart also made a number of recordings that were not released.

8. New Records for March 1915, Redpath Collection, box 319.

9. Taggart to Crotty, December 11, 1914, Redpath Collection, box 318.

10. Taggart prepared a sample letter to send to lyceum or Chautauqua committees in the communities where he was scheduled to appear, informing them as follows: "I am to entertain in your city soon. Now would it not be a good plan for you to see the dealers that handle the 'Victor Talking Machine' goods and see if they will get out a special ad for my records? It seems to me that it will boost the entertainment as well as the sale of records. I just offer this as a suggestion." Redpath Collection, box 319. From a letter from the Redpath Publicity Department to the Victor Talking Machine Company, October 22, 1917: "We are in receipt of a letter from the Redpath Publicity Department from C. R. Taggart stating that you will send us a quantity of circulars telling of him and his records. We will be pleased to have you send us 7,500.... We will see that the circulars are sent to the various town where he is booked." Redpath Collection, box 318.

11. Taggart to Harrison of Redpath, March 8, 1915, Redpath Collection, box 318.

12. Taggart to McClure of Redpath, December 15, 1915, Redpath Collection, box 319; emphasis in the original.

13. FM of the Redpath Bureau to Taggart, January 5, 1915, Redpath Collection, box 318.

14. Harrison of Redpath to Taggart, October 20, 1915, Redpath Collection, box 319.

15. N. S. Greene of Victor to Redpath Bureau, April 6, 1921, Redpath Collection, box 319.

16. Taggart to Harrison of Redpath, March 8, 1915, Redpath Collection, box 318.

17. H. P. Harrison to Redpath Managers, November 28, 1916, Redpath Collection, box 318.

18. Taggart to Crotty of Redpath, February 18, 1918, Redpath Collection, box 318.

19. Redpath Collection, box 319.

20. There is a vast literature on the Yankee as a characterological figure, well beyond the scope of this chapter. See Bryan 2013; Conforti 2001; Dorson 1946; Hrkach 1998; Morgan

1988; Nickels 1993; Rourke 1959. Gal and Irvine (2019, 138–153) offer a refreshingly original perspective from the vantage point of linguistic anthropology.

21. I draw in the following discussion of Taggart's dialect from Hughes 1959; Kurath 1939a, 1939b; and Wentworth 1944.

22. New Victor Records, March 1915, Redpath Collection, box 319.

23. John Jakle (1999, 3), in his important article "America's Small Town / Big City Dialectic," states, "Seventy percent of all Americans in 1880 lived in rural areas or in towns of fewer than 2,500 people; 42% of the labor force was engaged in farming. In 1880, some 25% of the nation's population resided in urban places larger than 5,000, climbing sharply to nearly 40% in 1900. Between 1900 and 1910, the nation's cities grew by some 12 million people—30% from migration off farms and from villages in the United States."

24. In larger scope, this concern with "the problem of country life" is closely related to the core argument of Walter Lippman's hugely influential book, *Public Opinion* (1922), in which he contended that the Founding Fathers' ideological construction of "the farming communities of Massachusetts and Virginia" as the "the image of what democracy was to be" was no longer adequate to the needs of a modern, complex polity. The incompetence of rural school boards gave the lie to the Founding Fathers' "doctrine of the omnicompetent citizen," based on their conception of "the isolated rural township" and its democratic governance.

25. Victor Records for February 1916, Redpath Collection, box 319.

26. Columbia A2890, mx 78343-4, June 17, 1919.

27. "Old Country Fiddler at the Dance," Victor 35632-a, June 23, 1916.

28. "Old Country Fiddler at the Wedding," Victor 35538-b, February 4, 1916.

29. The interrelationships linking language, music, and chronotope in the symbolic construction of contemporary "country" people has been richly explored in two ethnographic works: Aaron Fox's (2004) *Real Country* and Alexander Dent's (2009) *River of Tears*.

SIX

—⚏—

"I DON'T SEE NO MANS"

Bridging the Schizophonic Gap

> "Gracious!" exclaimed an old man.
> "I hear speeches, but I don't see no mans."

I draw my epigraph to this concluding chapter from a newspaper account describing a bystander's response to a recorded speech played from a wagon drawn through the Lower East Side of New York City during William Randolph Hearst's 1906 gubernatorial campaign.[1] The confusion experienced by our disoriented hearer encapsulates a key feature of recorded sound identified by theorists from Theodor Adorno to the present—namely, that recorded performances are disembodied and no longer "tied to their place and time" (Adorno 1990a, 58; 1990b, 54). This is the phenomenon termed "schizophonia" by R. Murray Schafer (1994, 88; cf. Harkness 2011), the "splitting of sounds from their original contexts." Schafer identifies two dimensions of schizophonic decontextualization: "We have split the sound from the makers of the sound" (1969, 43) and also from their original occurrence "at one time in one place only" (1994, 90). To be sure, our Lower East Sider adds still another aspect of decontextualization to the separation of sound from its maker and from its originary situation of production—that is, the sensory decoupling of the visual from the auditory, not being able to see the speechmaker as you hear him, as you would in a live oratorical performance. Thus, in terms of the relation between sensory modalities, we are concerned with the potential manipulation and comprehension of the relation between the voice, the ear, and the eye (cf. Ochoa Gautier 2014; Weidman 2015).

In the foregoing chapters, I have addressed aspects of schizophonia as a problem for performers on early commercial sound recordings. Our puzzled old man, though, reminds us that schizophonia was also a problem for audiences and thus involves what Steven Feld (2015, 15) has identified as "the relational practices of listening and sounding and their reflexive production of feedback." And, as such, it was a problem for the nascent recording industry in its efforts to build audiences for sound recording as a medium of home entertainment. Entertainment is fundamentally about the enhancement of experience, the provision of pleasure for an audience. As a form of entertainment, performance is evaluated in terms of its ability to engage and hold the attention of the audience and to fulfill its desire for pleasure and gratification. The key question for our purposes was whether an audience that was accustomed to the experiential and sensory fullness of co-present performance would be satisfied with entertainment that relied on mechanically produced sound alone. We have seen and heard early recording artists working to orient their potential audiences to this problem as a reflexive dimension of the performances themselves. What I want to bring to the fore in these closing pages, though, is the role played by early record performers off record, so to speak, in addressing this problem as well in recording-oriented terms. In particular, I highlight two principal strategies: supplementation by photography and "record-taking" demonstrations.

PHOTOGRAPHIC SUPPLEMENT

The simplest effort to compensate for the schizophonic effects of commercial sound recordings was directed at the stripping away of sensory access to the originary performance, leaving open the auditory channel alone. Recall our elderly East Sider who could *hear* the recorded speeches that sounded from the back of a passing wagon but could not *see* the man who was delivering them. Essentially the same problem—the separation of the ear from the voice—affected all listeners to commercial recordings on their home talking machines. Here is the editor of *Phonogram*, writing in June 1900: "I met Cal Stewart (Uncle Josh Weathersby) not long since and I asked him for his Photograph. Now if you've never seen Cal, you can only half appreciate his droll humor and his bubbling laugh."[2] The editor goes on to describe Stewart in some detail, filling in verbally those additional features of his appearance (in character) that would make full appreciation possible. It is the anticipated photograph, however, that will allow record owners to supplement the experience of hearing Cal by seeing him (or his likeness) as well. "If he don't forget about his promise I'll get his Photo, and . . . I promise to print the picture in some future issue."[3]

Charles Ross Taggart's photographic supplement was considerably more imaginative. In addition to his talents as a storyteller and a musician, Taggart

The Old Country Fiddler Hears His Own Voice.

Fig. 6.1. Publicity photograph by Charles Ross Taggart in his guise as the Old Country Fiddler, *The Old Country Fiddler Hears His Own Voice*, 1915.

was an accomplished amateur photographer, taking pictures of himself in character as the Man from Vermont and the Old Country Fiddler, as well as in other guises. Sometime around 1915, early in his recording career, he produced a photograph of himself, *The Old Country Fiddler Hears His Own Voice*, for use in promotional materials for his lyceum tours and advertisements for his Victor records.

The image portrays Taggart in his embodied platform guise as the Old Country Fiddler, slouch hat on his head, fiddle in his hand, toothless grin on his face. These are familiar characterizing features of his stage persona, which viewers are tacitly invited to project upon the performer whose voice emanates from their record player. In the photograph, they can now see the performer they hear on record. The action Taggart is engaged in, though, plays no part in his platform performance: he is pictured sitting in front of a high-end cabinet model Victor Victrola, a material component of his remediated performances, conveyed on Victor records. His head is inclined toward the talking machine, and he is cupping his ear to focus and enhance the recorded sound emanating from the Victrola. He is not just hearing his own voice; he is *listening* to it. This is attentive, focused, intensified hearing (Rice 2015, 99). He is, then, assuming the stance not of his platform audience but of his record audience, hearing only the performer's (his own) voice. Even further, he is mirroring the desired reaction of his record audience: amused by, enjoying, and even marveling at the vocal performance issuing from the Victrola. This reaction is conveyed even more explicitly in the alternative caption that accompanies the photograph in the ad announcing new Victor records for February 1916: "The Old Country Fiddler Tickled With Himself." In sum, the photograph portrays Taggart in his platform guise listening to himself in his remediated guise, on record. As we look at the photograph, we *see* him, as does his platform audience, as he *listens*, as does his record-playing audience. He is having the same experience that we do, as listeners. Rather than schizophonically splitting the listening record audience from the performing source of the recorded sound, it *merges* the audience and the source into a unified reflexive image.

"RECORD TAKING" AS PERFORMANCE

If the recording process decoupled sound from its maker, yet another way of bridging the schizophonic gap was to reunite them. One of the most effective techniques that Edison employed from the very beginning in publicizing his tinfoil phonograph was to offer demonstrations of the recording process before journalists, scientists, or interested visitors to his laboratory, speaking some words into the recording horn and then playing back the recording to the amazement and delight of the audience. The performance potential of recording demonstrations was quickly recognized by other adepts, and record-taking demonstrations became a popular form of entertainment well into the 1890s. The simulation of a campaign event discussed in chapter 2 was a case in point. With the rise of attention on the part of the industry to the cultivation of sound recording as a medium of home entertainment, the staging of record-taking demonstrations shifted attention from the skill of the recording technician as a technological adept to the virtuosity of the performer as the maker of popular recordings.

An early report of record-taking events featuring performers who were coming to be known as recording artists appeared in the trade magazine *Phonoscope*, produced by Russell Hunting, himself a popular recording artist and a featured performer in the events reported in his article.[4] The piece reports on "evening exhibitions of record-taking at the Columbia Phonograph Company's parlors which are drawing vast crowds nightly and are becoming immensely popular. The entertainments are under direction of Mr. Harry C. Spencer and the reproductions from the records taken of the various artists are so perfect as to evoke great enthusiasm on the part of the public." Spencer, like Hunting, was also an early recording artist, widely known for his Irish dialect performances. Hunting goes on to observe that "the parlor record-taking exhibition has indeed made a hit" and to predict that it "has evidently come to stay." Events like this one reunite the recorded sound with the makers of the sound, allowing the assembled audience to experience the living presence of the recorded voices they hear and—the promoters would hope—continue to hear when they play the artists' records at home. The record-taking exhibition becomes, at least for the audience members in attendance, the originary context—the time and place in which the sound and the makers of the sound are conjoined, eliding the schizophonic gap. The place, it should be noted, is the very site where the records were produced for the growing market—namely, the Columbia Phonograph Company's own studios, which constituted, in Michael Silverstein's (2004) apt phrase, a ritual center of semiosis, from which the recorded performances emanated throughout the growing mass market.

The outward reach of such record-making exhibitions, still pointing back to New York as the center, is nicely illustrated by an account, three years later, in the *Edison Phonograph Monthly*, an Edison house organ for its distributors.[5] Citing a report of the Merchants and Manufacturers Exposition at Milwaukee, the account notes,

> One of the most interesting features of the exposition is the making of Phonograph Records, in connection with the exhibit of the McGreal Bros., agents for the Edison line. It is an unusual opportunity to see two men whose voices have been made familiar through the medium of the Phonograph, while the men themselves were in New York. . . . Consequently, a large crowd surrounded Arthur Collins and Byron G. Harlan, who, with heads close together, sang into the mouth of an octagonal horn, and then stood by while the Records were placed in a Phonograph and reproduced. Both are men of splendid physique and strong voices, and their work is a practical demonstration of the statement that the Phonograph "listens as well as talks." Mr. Harlan and Mr. Collins are both employed in the Edison laboratory of New York, and are two of the most successful makers of Phonograph Records

in the country. Their duets and solos are reproduced throughout the country, Mr. Harlan having a tenor voice and Mr. Collins a baritone.

Mr. Collins, when asked how many Records he and Mr. Harlan had made, replied: "Millions of them. I have been in the business eight years and Mr. Harlan has been at it five or six. There are not many people who make a success of it, owing to the fact that it requires an iron throat, powerful lungs and a peculiar singing voice."

In this recounting of a record-making demonstration, the schizophonic gap that is a consequence of sound recording is clearly stated: Collins's and Harlan's disembodied *voices* have emanated from the distant center of the Edison laboratory and have become familiar to record buyers through the medium of the phonograph, while their *persons* have been in New York. Their recorded voices have been separated from the performers who produced them far from the many places throughout the country to which those voices have circulated on phonograph records. Now, here they are before a co-present audience in Milwaukee, manifested in all their vital physicality, with "splendid physique[s]," "strong voices," "iron throat[s]," and "powerful lungs." They are fully present in the act of recording and remain so while their just-produced records are played back to the very audience that has seen and heard them singing into the horn. Their recorded voices have become a feature of a situated live performance before an audience, in the very act of making a record. The schizophonic gap has been entertainingly elided, the recorded voices reunited with the physical sources of those voices in the situated act of making and playing back a record.

Even after sound recording was well established as an entertainment medium, record producers, performers, and dealers continued to stage events that featured live performers and their recordings for the growing public constituted by those who had already adopted the medium and those who might not yet have done so. The Edison Company, for instance, sponsored a continuing and popular series of "tone tests," held free of charge in large performance venues in which well-known Edison recording figures performed live before an audience a selection of pieces that they had also recorded. With the performer onstage was an Edison phonograph. In the course of the performance, the house lights were dimmed to the point of total darkness. After a short interval, the lights came up again to reveal the phonograph standing alone, carrying on the performance in the absence of the performing artist, thereby demonstrating the acoustic fidelity of the machines (Fabrizio and Paul 2002, 141; Sterne 2003, 261–264). These tone tests, in effect, played with the schizophonic dynamic of presence and absence, all in the interest of persuading the public that the technology of sound recording was eminently capable of serving as an effective medium of entertainment, that the schizophonia attendant on recorded sound would not compromise its ability to entertain an audience, even in the absence of the living performer.

Cal Stewart, always ready to capitalize on his fast-growing fame as a recording artist, was especially committed to making live appearances sponsored by talking machine and record dealers in which he demonstrated record making by performing and recording stories in his persona of Uncle Josh Weathersby. "See, Hear and Meet the man who made the 'Uncle Josh' Records of Yankee Stories famous the world over" (McNutt 2018, 51). These free events were meant to appeal both to those who already owned phonographs and records and to those potential consumers who might consider doing so. As one advertisement put it, "We take great pleasure in announcing to the general public as well as to the owners and users of Phonograph and Graphophone machines and records, that we have secured the services of CAL STEWART, the progenitor of the famous Yankee stories of UNCLE JOSH WEATHERSBY, for a free exhibition at our warerooms" (McNutt 2018, 56). An especially attractive feature of Stewart's exhibitions was that in his guise as Uncle Josh, he made "original," personalized records for individuals in attendance, which they were free to take home with them afterward (McNutt 2018, 10–11, 51). To frame these made-on-the-spot recordings as "original" is to complicate Schafer's (1969, 43) proposition that it was only before the advent of sound recording that "every sound was an original." Not only did these personalized records contain the names of fortunate individuals, they also varied in detail improvised on the spot from the familiar, commercially available versions of Stewart's stories. Consider the implications of this technique. As in the other cases we have considered, Stewart's live demonstrations bridged the schizophonic gap, conjoining the living voice, its embodied source, and the recorded voice. But the records that he produced on the occasion also displayed features of oral traditional storytelling in their adaptation to elements of the unique situational context in which they were recorded. For those who took home a record made for them in one of these live events, the recorded performance would index the live performance every time it was replayed.

This complex play of relationships with the dynamics of schizophonia also had implications for the dynamics of scale in the nascent mass medium of commercial recordings. A 1900 ad posted by an Akron, Ohio, dealer is especially suggestive in this regard. "A general invitation is extended to all," it says, "to become personally acquainted with the man that has made the millions

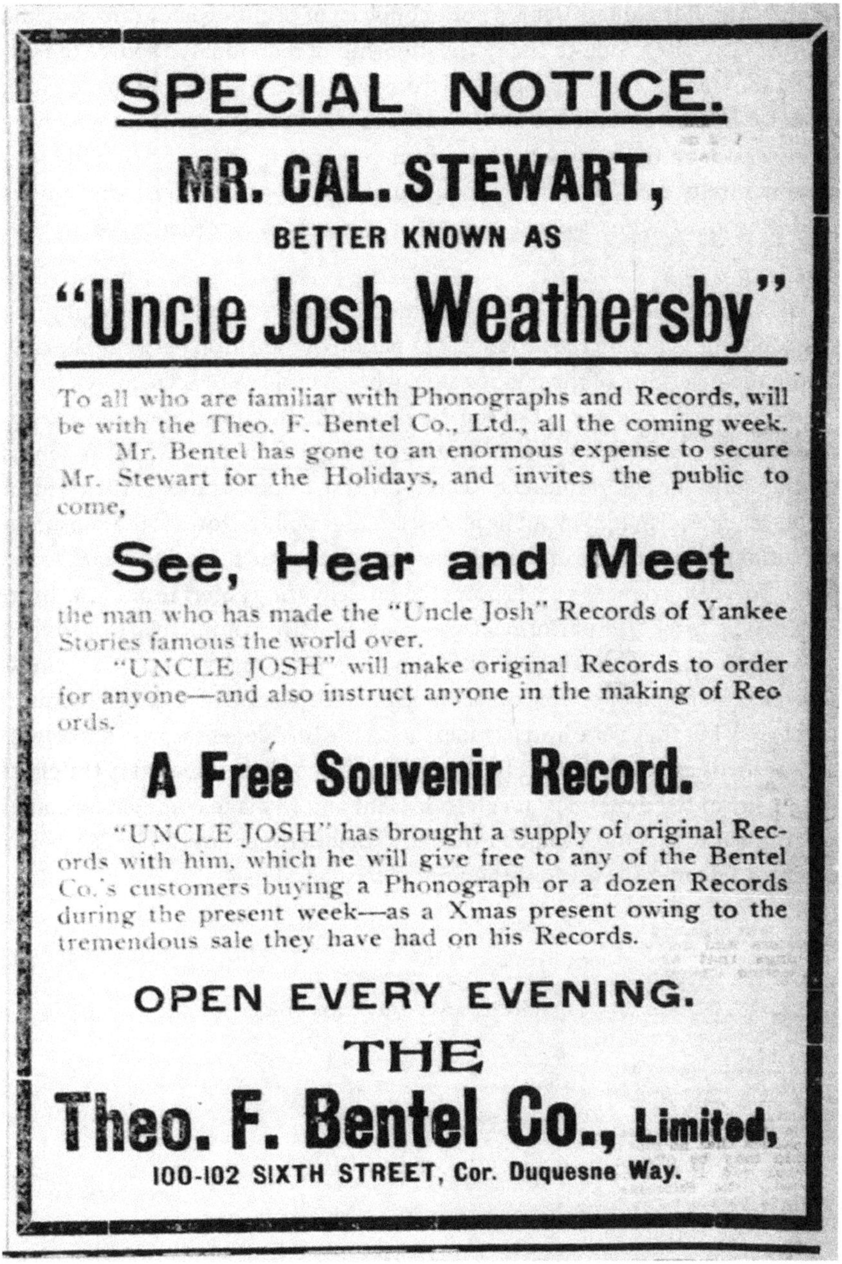

SPECIAL NOTICE.

MR. CAL. STEWART,

BETTER KNOWN AS

"Uncle Josh Weathersby"

To all who are familiar with Phonographs and Records, will be with the Theo. F. Bentel Co., Ltd., all the coming week.

Mr. Bentel has gone to an enormous expense to secure Mr. Stewart for the Holidays, and invites the public to come,

See, Hear and Meet

the man who has made the "Uncle Josh" Records of Yankee Stories famous the world over.

"UNCLE JOSH" will make original Records to order for anyone—and also instruct anyone in the making of Records.

A Free Souvenir Record.

"UNCLE JOSH" has brought a supply of original Records with him, which he will give free to any of the Bentel Co.'s customers buying a Phonograph or a dozen Records during the present week—as a Xmas present owing to the tremendous sale they have had on his Records.

OPEN EVERY EVENING.

THE

Theo. F. Bentel Co., Limited,

100-102 SIXTH STREET, Cor. Duquesne Way.

Fig. 6.2. Newspaper advertisement announcing a "record-taking" appearance by Cal Stewart, *Pittsburgh Daily Post-Sun*, December 16, 1900.

laugh" (McNutt 2018, 56). It is a commonplace of thinking about mass media that they involve a "one-to-many" relationship of individuated source to multiple anonymous receivers, separated from each other by time and space. Here, though, those in attendance at Stewart's record-making events are massively downscaled, for the moment at least, from one among anonymous millions to a named individual, personally acquainted and co-present with the famous recording artist, who is so widely familiar through his recorded voice.

—␉␉—

These various techniques and experimental efforts aimed at bridging the schizophonic gap were the products of a period of transition between the time before the invention of the phonograph, when people were accustomed to engaging performance as part of the co-present interaction order, and the approaching era in which sound recording would be fully established as a mass medium of home entertainment. Record-making demonstrations still relied on co-present performance in public places, but they pointed toward the consumption of sound recordings in the privacy of the home and helped to socialize the public to its affordances and its potential for delivering entertainment to them in domestic space. The performers who made the early commercial records had an obvious stake in making the medium as entertainingly successful as possible. Much of their effort, of course, went into the process of making records in the studio, but they were also participants in the broader enterprise of promoting the medium to the public. Both on and off record, they shared in the effort to transform the dynamics of entertainment and to make commercial sound recording into the "most valuable medium" that Berliner foresaw back when its commercial potential was only beginning to suggest itself.

NOTES

1. NYT, October 24, 1906, 1. (See discussion in ch. 1, p. 24–25.)
2. Shattuck 1900, 58, 62.
3. Shattuck 1900, 62.
4. "Our Tattler" 1898b, 9.
5. EPM, June 1905, 13.

DISCOGRAPHY

PATRICK FEASTER

A BOOK LIKE THIS ONE is expected to cite its documentary sources in ways that give readers the information they would need to locate them, access them, check them against whatever claims an author may have based on them or made about them, and possibly use them to inform new arguments in turn. But what does that mean when those sources happen to be commercially issued sound recordings from the late nineteenth and early twentieth centuries? Style guides tend to have comparatively little to say about citing sound recordings in general, and that little isn't well suited to handling material from the industry's formative period, when techniques for recording and duplication were still just being worked out, and methods of physical labeling and differentiation along with them. But we take early sound recordings seriously as objects of study and sources of evidence, and we hope to encourage others to do likewise. With that in mind, we want to try to provide citations for them that could really be acted on—as well as some practical guidance in *how* to act on them—and that aren't merely for show. So what kind of information is in fact needed to identify these sources uniquely—or, in other words, to distinguish a copy of the "same" recording from a copy of a different one? And how much does it matter?

One reason why readers might want the specific sound recordings discussed in this book to be clearly identified is, of course, that they'd like to try to track them down and listen to them in conjunction with what they're reading. Sound recordings can't be quoted directly as audio on the printed page, and textual transcriptions, no matter how meticulous, can't provide the replete listening experiences characteristic of the phonographic medium, which readers might understandably want to have for themselves. An author or publisher can sometimes sidestep this part of the problem by compiling the recordings into an audio companion piece—a compact disc tucked into an envelope on the back cover, for example, or a set of digital files hosted on a dedicated website (although it's not clear that either of these options will enjoy the same longevity as a printed book). This kind of supplementary audio shares a common rationale with such other forms of directly presented evidence as the quotation, the graphic illustration, and the appendix of source documents. But as in those other cases, directly incorporating a source doesn't obviate the need to identify its origin clearly and unambiguously. Indeed, its identification arguably becomes all the more urgent, since others are likely to tap it in turn as primary evidence and should know what it is they're using. After all, there's more at stake in these cases than merely being able to access the sources as such or to listen to recordings as recordings. There's also the matter of establishing what kinds of objects the sources are in the first place, which can have implications in turn not only for how someone might go about verifying their content in case of doubt but also for the nature of the conclusions someone might defensibly draw from them. Most readers are probably aware that many books exist in multiple editions, that works of ancient and medieval literature often exist in multiple manuscripts, that films can exist in multiple cuts, and that what's true of one edition, manuscript, or cut may not be true of others. It may be less obvious, or less well known, that early commercial sound recordings were similarly variable. But they were, in many cases, and some of the recordings examined in this book are better regarded as single manifestations of works that exist in other variants as well, like one telling of a story among many.

One of the first selections quoted in the introduction to this book is "Street Fakir," in which George Graham assumes the role of a pitchman touting the wonders of Doctor Boccaccio's Celebrated Egyptian Liniment. Here's another passage from it that draws in another of this book's themes—presidential politics—although it isn't especially noteworthy from the individual standpoint of either politics or medicine shows and so didn't make the cut when those topics were being considered separately before.

> There was Grover Cleveland,
> was laid up with the rheumatism so that he could not move,
> couldn't walk.
> He used one-quarter of a bottle of this preparation,
> and today he is as well as ever
> and is in good shape to walk out o' town
> the next fourth o' March.

The source for both this quotation and the earlier ones is a seven-inch single-faced shellac disc manufactured for use on Emile Berliner's gramophone, the direct technological ancestor of the modern turntable. Gramophone discs from this period don't have paper labels, as later discs do; instead, written information about the title, performer, and so forth was molded into pressings from the original master plates along with the groove itself, often scratched in Berliner's own crabbed cursive. The date March 23, 1896—cited in the introduction—had been included among this handwritten metadata, as it often was, which is how we know so precisely when the underlying performance was recorded. Moreover, we know on the same basis, from an examination of surviving pressings, that Graham performed "Street Fakir" not just once for Berliner's gramophone that day but at least twice. The text quoted here and in the introduction comes from a disc stamped "638Y." That number, 638, was the public-facing catalog number used for submitting and filling orders, and it refers to "Street Fakir" as a selection in general, irrespective of performer. Meanwhile, the Y suffix was intended strictly for the company's own internal use and designates a unique master recording, typically referred to today as either a "take," in reference to its performance, or a "matrix," in reference to the physical plate containing it. With Berliner discs from this period, the usual protocol—inferred by collectors studying many actual discs over many years—was for the first take of each numbered selection to have no letter, while subsequent takes received letters in reverse alphabetical order starting at Z. In this case, pressings exist of both 638Y and 638 (with no letter), both bearing the same hand-scrawled date and hence known to have been recorded on the same day. If you happened to read along in the transcription while listening to the latter take, you might notice a few discrepancies: Graham says "could not walk" instead of "couldn't walk" and "now he's as well as ever" instead of "today he is as well as ever," and he doesn't pause between the last two lines quoted above. If you didn't know better, you might chalk these discrepancies up to errors in transcription when, in fact, they represent legitimate differences among takes. As it is, a comparison reveals that Graham's wording was fairly consistent from performance to performance at the time but not rigidly fixed; he was plainly not reading from a script but reciting a memorized piece, such that some variation could creep in from iteration to iteration. Otherwise, the two takes are largely interchangeable for purposes of analysis. Barring minor differences that impinge on verbatim transcription but are otherwise scarcely noticeable, nearly all the observations someone might make about 638Y would apply just as well to 638.

That said, it would be premature for us to conclude on the basis of the two takes we've considered so far that George Graham's "Street Fakir," as such, contains a segment about President Grover Cleveland being cured of rheumatism. Both takes were recorded on the same day and represent a distinctive historic moment when "the next fourth o' March" was March 4, 1897: the specific date when Cleveland, who wasn't seeking reelection, was scheduled to leave office. In that context, the Cleveland segment was both topical and clever. The hand-scrawled metadata visible on 638Y include not only a date but also a place: "WDC," meaning Washington, DC, where Berliner maintained a recording laboratory, so that "out o' town" was also anchored deictically to the city in which the originary performance happened to be taking place—the same city Cleveland would leave at the end of his term and the one in which Graham himself was most accustomed to performing on street and stage.

But Graham also performed "Street Fakir" for the talking machine on subsequent occasions when the Cleveland segment would no longer have been so timely. In one case, we find him substituting a reference to a different political figure: Senator Mark Hanna of Ohio, chair of the Republican National Committee.

> Now there was Mark Hanna,
> was so laid up with the rheumatism that he could hardly walk.
> He used one half bottle of this preparation
> and today is able to walk around
> and collect money
> for to help to elect McKinley.

The source this time is a seven-inch single-faced Zonophone disc—the product of some of Emile Berliner's erstwhile business partners who had driven him out of the American market in 1900 and introduced discs and machines of their own make. It's stamped "V9304," with 9304 being the catalog number and the V serving as a genre marker for spoken-word selections. A 3 handwritten at twelve o'clock and a 2 stamped between two and three o'clock may be take markers, but such details are less well understood for Zonophone than for Berliner. There's no written date this time, but we can make an educated guess as to when it was recorded by drawing on less direct sources of evidence: the physical characteristics of the disc, including the presence of a shield logo known to have been used for only a limited time; lists of records found in dated catalogs; the position of the number 9304 relative to other sequentially numbered releases for which dates can be estimated; and the recorded content itself, which in this case implies that, as of "today," the presidential election of November 6, 1900, lay in the near future. Taken together, all these indications converge on a date shortly before November 1900.

The segments about Grover Cleveland and Mark Hanna differ in their specifics, but they follow a consistent formula. The pitchman's liniment is supposed to have made it possible for a rheumatic politician to walk again in general, enabling him to do some more specific kind of walking associated with the vicissitudes of political life: Cleveland could walk out of town at the end of his presidential term, and Hanna could walk around raising funds for William McKinley's reelection campaign. Judging from these two variants, we might hypothesize that this formula, at least, was a consistent part of Graham's "Street Fakir" and that he regularly updated it, much as he might have done for live medicine shows, pairing the name of some current politician with an associated activity that involved walking. However, other takes reveal an even broader range of variation. In a different version on Berliner gramophone disc—this one dated October 11, 1899, with catalog number 0585 and no take marker—we find the following segment at the usual point in the routine:

Now there's James J. Corbett,
the noted pugilist.
He was laid up with the rheumatism so that he could not talk.
He used one half a bottle of this preparation,
and now he can talk so much
that he fills up four columns in the newspaper
telling about how he will win the championship
next year.

The subject in this case is a professional boxer rather than a politician but still, as before, a prominent public figure: at the time this recording was made, Corbett was eager to reclaim a heavyweight title he had lost back in 1897, and a much-publicized championship match with Jim Jeffries was arranged for "next year" shortly thereafter. Rheumatism is supposed here to have prevented Corbett from talking rather than walking—hoarseness is a comparatively rare but possible symptom of rheumatoid arthritis—but still, as before, the activity in question was central to the maintenance of his role as a public figure.

In 1903, Graham reprised his "Street Fakir" routine yet again for the Victor Talking Machine Company. This case differs in several particulars from the ones we've considered so far. Any original ledgers associated with the recording of Berliner or early Zonophone discs have long since been lost, so our knowledge of what was recorded and when is limited to what can be gleaned from catalogs and from the discs themselves. By contrast, Victor's recording ledgers survive, as a result of which their recordings tend to be documented with unusual thoroughness and detail—not only issued recordings but also takes that were rejected and never put into commercial production. Victor discs have paper labels, unlike their predecessors, and the title of Graham's routine appears printed on them not as "Street Fakir," but as "Imitation of a

Street Fakir," which is also how all the earlier takes had been introduced in their spoken announcements, such that the same title could arguably apply to them as well, depending on whether we consider written or spoken titles to be authoritative. Victor assigned this selection the catalog number 2617 and offered it in two different sizes, measured by diameter: as a seven-inch disc, like the Berliner and Zonophone discs considered above, and as a ten-inch disc with the same number, marketed under the name Monarch Record. The usual practice is to distinguish Victor recordings of different sizes by prefixing letters to their numbers, such as V-2617 and M-2617, where V stands for Victor (seven inch) and M for Monarch (ten inch). Multiple takes can be associated with each size in turn, often recorded on different days, sometimes years apart. Victor's practice was to leave the first take physically unmarked but to mark each subsequent take with an ascending number starting at 2, adjacent to the label area at nine o'clock, where it is often but not always visible on pressings. According to the ledgers, Graham produced first takes of "Imitation of a Street Fakir" in both sizes on April 9, 1903, but it was the second takes—V-2617-2 and M-2617-2, recorded during a follow-up visit on April 22—that were put into commercial production. We know both dates only thanks to the survival of the ledgers. On the earliest Victor discs, handwritten dates can sometimes be read pressed from stampers into the shellac underneath the paper labels, much as they can be read on Berliner discs without paper labels; but from 1903 onward, the central area of Victor pressings was recessed to protect labels against scuffing, thereby expunging whatever internal metadata might have been scrawled there on the master plates. As a result, the dates can no longer be discerned just by examining the discs themselves and always need to be looked up in a source that draws on the ledgers, the standard reference work of this kind being the *Encyclopedic Discography of Victor Recordings* (EDVR). The first installments of EDVR came out in book form, but the project moved online in 2008 and has since been subsumed into the online Discography of American Historical Recordings—DAHR, at adp.library.ucsb.edu—a database that contains listings for numerous other companies as well and has become an indispensable resource for anyone working with this kind of material.[1]

Coarse-groove shellac gramophone discs may now be known colloquially as "seventy-eights," but during the 1890s and early 1900s, recording speeds hadn't yet become standardized at 78 rpm and tended instead to span a range extending downward into the 60s. Consequently, there's always some degree of guesswork involved in setting playback speeds for these recordings today. But even bearing that uncertainty in mind, ten-inch discs generally had more capacity than seven-inch discs, such that when Graham performed "Street Fakir" for ten-inch Victor M-2617-2, he had more time to fill than before, permitting—or perhaps even demanding—more elaboration of the routine. While the Cleveland, Hanna, and Corbett segments quoted above average about fifty words each, the equivalent segment on M-2617-2 clocks in at a significantly longer seventy-two words. The subject is once again a politician and, like Cleveland in 1896, a sitting president:

> Now there's President Roosevelt,
> he didn't have any voice,
> his voice gave away.
> He gargled his throat
> with one-sixteenth of a grain of this—medicine,
> and now his voice is heard
> from ocean to ocean.
> And he was troubled with bad teeth,
> and he used some of this preparation on his teeth,
> and his teeth can be seen the same distance
> with the naked eye.
> What do you say to that?

Like Corbett, Roosevelt is supposed to have lost his voice, although not necessarily to rheumatism, but Graham takes advantage of the extra time he has available to add a second claim about bad teeth, perhaps inspired by the popular association of Roosevelt with a distinctively toothy smile. Where the medicine had enabled Corbett to get his talk into the newspaper, its alleged effect on Roosevelt's voice and teeth is more direct: both could now be perceived "from ocean to ocean."

Given our interest here in the phonographic remediation of speaking, this latest set of claims might well attract our attention. After all, the phonograph too was known for extending the range of people's voices "from ocean to ocean," and although it didn't do this specifically for Roosevelt's voice until 1912, it had been doing it to Graham's own voice for some years by the time he performed "Street Fakir" for Victor. Meanwhile, we know that Graham had personally experienced a condition at least something like the ones he imputed to Cleveland, Hanna, and the rest. He himself suffered from rheumatism—so much so, in fact, that his fellow Washington entertainers held a charity benefit for him in 1895, implying that his affliction must have been severe enough at times to impede his ability to earn a living from his verbal art.[2] Medicines like

Doctor Boccaccio's probably didn't give him whatever relief they promised, lending some poignancy to his "imitations" of the pitches associated with them. But 1895, the year of that charity benefit, was also the year in which Graham appears to have begun working for Berliner's gramophone. He might have seen this, among other things, as an opportunity to earn some income even if he were, as he said of Cleveland, "laid up with the rheumatism so that he could not move, couldn't walk." And that prospect may not have been purely hypothetical. An engineer for Victor recalled one time when Graham had showed up to a recording session "unable to stand in front of the recording horn without assistance" and that he had built an iron yoke to hold Graham's forehead in place "so he could not wobble all about" while talking. The engineer ascribed Graham's condition to "the influence of liquor," but whatever its cause, the outcome would still have been to enable Graham's recorded speech to travel across the country in spite of his physical indisposition.[3] Considered from that angle, the testimonials Graham presents for Doctor Boccaccio's liniment arguably share something in common with his own experiences with the gramophone. Both involved lifting the constraints a physical ailment had imposed on someone's activities—activities that usually required a combination of mobility and speaking.

In terms of its variability per se, "Street Fakir" is typical of phonographic routines of the 1890s and early 1900s. During that period, a selection couldn't yet be recorded just once and then duplicated indefinitely to fill a steady demand, month after month and year after year. In the case of Emile Berliner's gramophone discs, only a limited number of duplicates could be obtained from each master plate before it wore out, after which another master would need to be pulled out of storage or recorded as a replacement if the selection were to be kept in production. Later, as new companies such as Zonophone and Victor entered the field, they each proceeded to record their own independent versions of whatever established selections they wanted to include in their own catalogs, frequently employing the same performers. Meanwhile, as recording methods evolved and new formats were introduced—such as the ten-inch disc—older selections were routinely "remade" to bring them up to current standards and to take advantage of technical innovations. Under these conditions, the more interest there was in commercially exploiting any given selection, the more different takes of it tended to circulate, with each of those takes derived from a separate performance and hence differing at least slightly from all other takes.

Those differences can be more or less consequential. Each take of "Street Fakir" contains a sequence of two testimonials: one that varies quite a lot from take to take, as we've seen, and another that usually comes second in the sequence and is far more consistent. But even this other testimonial displays some variation. On Berliner 638Y, it runs as follows:

> There was an old lady down in South Washington
> a hundred and seventy-five years old,
> had been bedridden for seventeen years.
> She used one half a bottle of this preparation
> and today she is earning a good living for herself
> dancing in the ballet in *The Black Crook*.

On Victor M-2617-2, the corresponding testimonial stands out for coming in first place, before the Roosevelt testimonial—reversing the usual order—but is otherwise quite similar:

> Now there was an old lady down in South Washington
> a hundred and seventy-five years old.
> She had been bedridden—for seventy-five years.
> She took one half bottle of this preparation
> and today she's earning a good living for herself
> dancing in the ballet—of *The Black Crook*.

The most notable change we find if we compare variants of this other testimonial is a ratcheting up of the hyperbole over time: the subject has been bedridden for just "seventeen years" on takes recorded through 1899 but for a marginally more impressive "seventy-five years" in and after 1900. On Zonophone V9304, she also takes a whole bottle of medicine, and not just half a bottle. Such minor differences in wording and detail affect most early phonographic spoken-word selections that exist in multiple takes, except for the ones conceived as recitations of standard texts such as Lincoln's Gettysburg Address. Takes recorded on the same day, such as Berliner 638 and 638Y, are typically more similar than takes recorded weeks or years apart. These differences are worth bearing in mind when we consider either the nature of these phonograms as sources or the dynamics of the performances that created them. At the same time, few arguments are likely to stand or fall based on the difference in this case between "seventeen years" and "seventy-five years." For most purposes, the variants of this testimonial will be functionally interchangeable.

But that's plainly not true of the other testimonial—the one that deals in different instances with Cleveland, Hanna, Corbett, and Roosevelt. And while "Street Fakir" is typical among early phonographic spoken-word selections for its variability per se, it's quite unusual—perhaps even unique—for drawing in different topics of the day with such regularity. What can we make of this? Graham was likely accustomed to working allusions to current and local events into his live performances, taking advantage of the fact that he and his audiences shared a common "now" and a common "here," the latter usually in Washington, DC. He may simply have carried that habit over into his work for the gramophone without factoring in the less predictable circumstances of future playback or even the changes in his own location as he visited different recording studios. As to the latter, Graham consistently introduces the two testimonials in "Street Fakir" as coming from prominent citizens "of this city" or, on M-2617-2, "of your city," even though one of them explicitly concerns a woman of South Washington, while the Zonophone recording studio was located in New York City and that of Victor in Philadelphia. The references to "the next fourth o' March" or "next year" might reflect a similar stance—one in which Graham aimed to speak into the recording horn precisely what he would have spoken to a live audience at the same moment in Washington. It's also possible that Graham and his employers expected each of his recordings to have a commercial life span of only a few months, such that they wouldn't meaningfully outlive the currency of the topical references in them. (They did in practice, though. After Berliner had been forced out of the talking machine market in the United States in 1900, he moved his operations to Montreal, where he continued to press fresh copies of some of his older matrices, including 638Y from 1896, which can be found bearing a telltale "MADE IN CANADA" patent stamp.[4] One wonders what Canadian audiences of the early twentieth century made of Graham's line about Grover Cleveland being "in good shape to walk out o' town the next fourth o' March.") Meanwhile, Graham's understanding of the rhetorical function of the Cleveland-Hanna-Corbett-Roosevelt segment may have obliged him to update it regularly with references to current events even if other factors, such as the guarantee of rapid obsolescence, would have weighed against him doing so. What made each of these testimonials so comically impressive was that it involved someone who was not only in the news but in the news specifically for an activity—usually involving virtuosic feats of speech and movement—that Doctor Boccaccio's medicine could be touted as having made possible. Cleveland fits the pattern least well, since Graham directly credits the medicine only with enabling him to "walk out o' town," or to leave office; but even he, as president, was associated more generally with moving about and talking to people in ways that severe rheumatism would have complicated. If Graham's intention in this part of "Street Fakir" was always to stake a claim about someone who was prominently in the news, that would explain its constant variation while the rest of the routine remained comparatively stable.

I've gone into "Street Fakir" in some detail here to draw out several points. First, distinguishing among versions of early commercial sound recordings isn't a sign of obsessive pedantry but a matter of responsibly managing differences among sources that can be quite significant. The comparisons, contrasts, and generalizations I laid out above, for whatever they're worth, obviously wouldn't have been possible without it. But even if someone were merely to write about, say, the testimonial involving President Roosevelt's voice being "heard from ocean to ocean," it would still be important to identify its source as Victor M-2617-2, not just as George Graham's "Street Fakir," because most versions of "Street Fakir" don't contain this passage and were, for that matter, recorded before Roosevelt had even become president. Second, the entity we want to identify is generally the take rather than the catalog number, since multiple takes were sometimes pressed under the same catalog number—witness 638 and 638Y—and although I haven't yet brought up an example of this here, the same take can also be found pressed under different catalog numbers. Third, the physical markings we can use to differentiate among takes vary from company to company and period to period, as do the notations commonly used to refer to them. Thus, in 638Y the take marker is the Y, while in 638 it's the absence of a letter, and in M-2617-2, it's the 2 visible on pressings at nine o'clock. Meanwhile, on Zonophone 9304, it's unclear whether it's the handwritten 3, the stamped 2, both of these numbers together, or neither of them; given the current state of knowledge, this case is legitimately and frustratingly ambiguous. Fourth, changes in the duration of recording media tend to coincide predictably with expansions and contractions of material, as illustrated here by the different lengths of Graham's testimonials on seven-inch and ten-inch discs. Fifth, the dates associated with early commercial sound recordings are typically dates of recording rather than dates of publication, perhaps because scholars—for example, of operatic singing or jazz—have been more interested in the underlying performances than in the circumstances of release; but at the same time, the recording date is usually not a handy point of reference we can use practically to identify an object in hand, as it commonly is with the publication date of a book. There are exceptions to this rule, such as Berliner gramophone discs with handwritten dates molded visibly into their label areas. But in most cases, recording dates need to be looked up or inferred, and depending on what ancillary sources of information are available, they can vary in precision from an exact day to a range of months or even years. The author-date citation style is particularly ill equipped to handle such material, but citing early commercial sound recordings in any style that foregrounds dates as an organizing principle always feels rather like forcing square pegs into round holes. Of course, there are plenty of reasons to want to know the date of a recording, but for purposes of identification, the date won't routinely have the practical value it has for distinguishing among, say, editions of books or issues of journals in which publication dates are commonly displayed and easy to check. When dates do appear on a disc label, they're usually patent dates, not dates of recording or publication, from which they often differ by a decade or more.

George Graham's "Street Fakir" was marketed not only on gramophone discs or "seventy-eights" but also on phonograph cylinders—as Columbia 10504—and the cylinder format introduces some unique complications of its own. At first, cylinder records weren't duplicated by pressing or molding, as gramophone discs were, but had instead to be cut individually onto blanks made of "brown wax," as the distinctive metallic-soap material used for them is called. Often that meant cutting records for sale directly from performances, a process that could be made more efficient by grouping together several phonographs so that they could all record a single performance at once. Alternatively, a master recording on one cylinder could be copied onto other cylinders for sale, either by playing it through a tube or by using a mechanical linkage known as a pantograph, but master recordings wore out quickly either way. Because these early duplicate cylinder recordings were copied by cutting them individually onto blank media, they have none of the molded-in physical markings we can use to identify different takes on disc. In most cases, the cylinders themselves are physically unlabeled, and their contents were originally identified in writing only by inscriptions on the lids of their containers or by inserted paper title slips, both of which have frequently gone missing. As a result, brown wax cylinders commonly have to be identified today on the basis of their spoken announcements—not only as to title, performer, and company but also as to date, based on verbal formulas that are understood to have varied over time: for example, "Columbia Phonograph Company of Washington, DC" points to 1896 or earlier; "Columbia Phonograph Company of New York City," to around 1897; and "Columbia Phonograph Company of New York and Paris," to around 1898–1899. There's nothing on brown wax cylinders equivalent to standardized take identifiers like the Y in 638Y, but even if there were, it would be of limited value: considering constraints on duplication and poor rates of survival, any cylinder from this period—even if it's a duplicate—is likely to contain the sole surviving record of whatever performance is captured on it. When it comes to identification and citation, brown wax cylinders are less like printed editions than they are like manuscripts, each one presumptively unique. To identify a specific brown wax source reliably, then, it's necessary to cite either a specific physical cylinder or some specific digital file, transfer, or reissue derived from it.

Let's consider one concrete example for which audio is readily available. The Columbia Phonograph Company offered "Lincoln's Speech at Gettysburg" in its catalog of cylinder records starting in 1895, initially under the number 3028, and from 1896 onward—after the adoption of a new numbering scheme—under the number 11004.[5] We have no way of knowing how many different times Columbia recorded it during this period or even who might have performed it for them, since the selection was listed anonymously, although Russell Hunting was frequently responsible. The online University of California, Santa Barbara (UCSB) Cylinder Audio Archive (http://cylinders.library.ucsb.edu/) has audio for two different cylinders identified as copies of Columbia 11004, both sourced from the John Levin Collection. Each item in this audio archive has been assigned a unique sequential reference number, and the items in this case are numbered 13669 and 16989. The one numbered 13669 is announced "Abraham Lincoln's greatest and world-famed address at Gettysburg, November eighteenth, eighteen sixty-three," while the one numbered 16989 launches straight into the address itself with no announcement—quite a significant difference in framing. Other copies of Columbia 11004 recorded and sold at the time might have resembled one or the other of these, but they might also have been announced in any number of other ways. And that's presuming that the identification of these two items with Columbia 11004 is correct. It's unclear from the metadata presented on the UCSB site what evidence—such as, say, a title slip—supports an identification of either cylinder with the catalog number 11004 as opposed to the earlier number 3028 or with Columbia itself as opposed to some other recording company. After all, Lincoln's Gettysburg Address was also available on brown wax cylinder as Edison Record number 3821; and while Columbia and Edison were the leading producers of cylinder recordings during the late 1890s, there were countless other, smaller-scale manufacturers besides, any of which might also have cut its own indistinguishable recordings of the same selection, drawing on the same small pool of specialist performers.

The number of copies that could be made from any one master recording increased markedly about the year 1902. For phonograph cylinders, Columbia and Edison introduced superior molded duplicates of "black wax"; going forward, these would display molded-in markings comparable to the ones seen on discs. Meanwhile, for cylinders and discs alike, a multi-stage duplication process was adopted as well. With discs, a master would now be used to mold a negative "father," which would be used to mold multiple positive "mothers," each of which could be used in turn to mold multiple negative stampers from which positive copies could be pressed for sale. Cylinders were handled similarly. Thanks to this precaution, the pace at which popular selections had to be "remade" steadily decreased until by the 1910s it wasn't unusual for even a good-selling record to exist in just one single issued take. The recordings discussed in this book straddle both sides of the transition.

As the process of duplicating records grew more complex, so did schemes for organizing and numbering them, which has implications in turn for the forms citations need to take. At first, Victor had used a single numbering system for both its catalog numbers and its own internal organization of matrices corresponding to specific master recordings, but in 1903 they began assigning separate numbers to their matrices, independent from the catalog numbering; recordings from before the change are referred to as "pre-matrix." On single-faced pressings, the matrix numbers are often visible handwritten at six o'clock and have a letter prefix corresponding to disc size: A for seven-inch, B for ten-inch, C for twelve-inch, D for fourteen-inch, and E for eight-inch. A single-faced Victor disc from this period will thus have both a catalog number and a matrix number. However, most Victor discs issued from 1908 onward were not single-faced but double-faced, with a different recording pressed onto each side, distinguished by a letter suffix. The two sides of Victor 18587, for example, are labeled 18587-A and 18587-B, each with a different matrix number corresponding to the master recording pressed onto it (B-21405-3 and B-22895-4, respectively). Matrix numbers aren't displayed visibly on double-faced Victor discs (or on later single-faced ones), but the matrix number for a given recording can still be looked up by its catalog number in EDVR, and the take number visible at nine o'clock can then point to a recording date in turn. Often the same matrices that had been used for single-faced pressings were later "coupled" to produce double-faced pressings. For example, Victor matrix B-5081-1, "New Parson at Darktown Church," was first used for single-faced pressings with the catalog number 5402 but later for the A side of double-faced pressings with the catalog number 16816. In neither of these cases is the matrix number displayed anywhere on the discs themselves.

During the decade of the 1900s, Victor's main competitor in the disc market was Columbia—a company that was unusual at the time for simultaneously producing both cylinders and discs. Unlike Victor, which came to distinguish single-faced disc catalog numbers from matrix numbers, Columbia used a single number for both purposes.[6] In 1908, Columbia—like Victor—converted most of its existing disc catalog to double-faced pressings, but it continued to print the single-faced numbers on labels in smaller type underneath the new double-faced catalog numbers, which have distinctive letter prefixes (all examples discussed in this book fall into the A series). After this point, Columbia continued to use its older numbering scheme for matrix identification, with the numbers in question appearing both on labels and molded into the shellac. Different takes were distinguished by number as well, but the take number appears only in the shellac, either separated from the selection's general reference number by a hyphen (e.g., 160-8) or, later, stamped above it. The most authoritative discography for Columbia's disc output, which has been absorbed into DAHR, originally appeared in print as a multivolume set in 1999; but in the absence of surviving recording ledgers, the authors had to reconstruct data from other evidence that is spottier for earlier periods than for later ones. When they cite specific dates, these generally reflect not when masters were recorded but when company records show they were shipped out for processing (Brooks and Rust 1999). For recordings made before the 1910s, the available information becomes even sparser. The date when Columbia first recorded a given selection can be cautiously estimated from when it debuted in catalogs, but there can be great uncertainty as to the timing of any subsequent takes or "remakes." A case in point is "Address of the Late President McKinley at the Pan American Exposition," originally issued as single-faced disc 833. This selection was first advertised in 1902, and the most common take is 833-10, which was used to press both single-faced discs and, starting in 1908, copies of double-faced disc A278. Based on that information, though, take 10 could conceivably have been recorded at any time between 1902 and 1908, and narrowing down its record-ing date much further would take a considerable effort (such as a survey of pressings, their physical characteristics, and their label types). Meanwhile, the performer is named neither on the label nor in a spoken announcement and can be identified as the prolific pioneer recording artist Harry Spencer only on the basis of voice recognition. For this period, the information we have avail-able about Columbia recording sessions is categorically less solid than it is for Victor recording sessions.

Even identifying a disc as a Columbia product in the first place isn't a trivial task. American pressings of Victor matrices generally display the name Victor or the "His Master's Voice" trademark—or both—on their labels, except for the very first discs pressed during 1900, before the company had settled on a permanent name.[7] By contrast, pressings of Columbia matrices can be found labeled with many other brand names, including Climax, Harvard, Oxford, Standard, United, Harmony, Busy Bee, D&R, Silvertone, Aretino, Star, and Lakeside, most of these being "client labels" marketed by third parties, such as mail-order companies. Columbia matrices recorded well into the 1910s can be found as single-faced pressings on client labels even if they only appeared double-faced in Columbia's own catalog, and double-faced client label pressings sometimes feature their own unique couplings. Columbia's distinctive-looking matrix numbers are usually visible in the shellac on client label pressings, but sometimes these were suppressed and replaced with unrelated control num-bers. Thus, the same Columbia matrix can often be found branded in a variety of ways and sometimes with its original identifying markings effaced. It's not difficult to recognize these "disguised" Columbia discs with practice, but the uninitiated might easily confuse them with discs that were in fact recorded by other, smaller concerns during the same period. Zonophone, for example, continued to maintain its own separate recording program even after it was acquired by Victor, while another company, Leeds and Catlin, pressed its own recordings under a variety of label names including Leeds, Imperial, Concert, and Sun. The products of these and a few other independent companies feature distinct recordings that can differ more or less significantly from those produced by Victor and Columbia. Complicating matters yet further, some client labels were sourced from more than one company. The Busy Bee label, for example, appears on pressings from Columbia, Leeds and Catlin, Zonophone, and the American Record Company, which can be distinguished from one another only through recognition of each company's distinctive numberings, typefaces, and so forth. On Oxford—a client label sold by Sears, Roebuck, and Company—discs of McKinley's last speech at the Pan-American Exposition numbered 833 were recorded by Columbia, whereas copies numbered 11849 were recorded by Leeds and Catlin. Meanwhile, for purposes of identifying one specific take, it doesn't matter whether a Columbia matrix such as 833-10 is found with a Columbia label or some other label, such as Oxford, Silvertone, or Standard, or whether it appears on a single-faced or double-faced pressing. The content will be identical in each case, although Columbia occasionally removed a spoken announcement from later stampers.[8] Our source for Columbia matrix 19236-2, "Congressman Filkin's Homecoming," happens to be a disc in my collection pressed with a "United Record" client label, but that detail would be of scarcely any use to anyone looking to make practical use of a citation.

Brown wax cylinders had even more variable recording speeds than contemporaneous discs did, extending from around 80 to 180 rpm, with spoken-word selections dominating the lower end of that range and with a general upward tendency over time (120–125 rpm was typical in the 1890s, while 144 rpm became more common just after the turn of the century). Once molded duplicates became the norm, however, 160 rpm emerged as the standard speed for commercial entertainment recordings. At that speed, standard cylinders—four inches in length, cut at 100 turns per inch (tpi)—could comfortably hold just under two and a half minutes' worth of content, a limit that the industry eventually took steps to increase. Columbia briefly marketed a longer-playing six-inch cylinder, but Edison launched a more enduring innovation in 1908 with the Amberol cylinder, doubling the available playing time to a little under five minutes by narrowing the groove pitch to 200 tpi. Rounding down, the two pitch standards came to be known at the time as "two-minute" (100 tpi) and "four-minute" (200 tpi). Each change that affected the maximum capacity of cylinders had a reciprocal impact on phonographic routines, which performers often expanded or contracted to fit the available time. For instance, Cal Stewart's 200 tpi cylinder version of "Jim Lawson's Horse Trade" runs about a minute longer than the versions found on 100 tpi cylinders at the same recording speed.[9] In another, separate development, black wax cylinders were superseded by cylinders molded from more durable celluloid, a material first introduced by such minor companies as Lambert but later adopted by Edison (the "Blue Amberol") and Columbia (which ultimately discontinued its own cylinder line in favor of marketing Albany Indestructible products). In 1912, Edison also introduced its own line of discs—albeit in a special format not compatible with standard gramophones—and some later cylinders were dubbed from disc masters so that the discs and cylinders contain identical recordings. Columbia's molded cylinders display take numbers much as Columbia discs do, but takes on Edison cylinders were differentiated in other, less intuitively obvious ways, such as by the quantity of dots following the inscription "PAT'D."[10]

So far, I've focused on the challenge of distinguishing one version or take of a selection from others. The necessary information isn't negotiable: it will either be present, or it won't. Beyond this, however, discographies—like bibliographies—can serve different purposes and can take different forms with those purposes in mind. (Note that the equivalent to a bibliography for sound recordings is conventionally known as a "discography" even when it lists objects other than discs, such as phonograph cylinders or magnetic tapes.) Some discographies, such as the ones incorporated into DAHR, aim to list the total activity of a given recording program. Meanwhile, other discographies try to list all recordings that fit criteria of some other kind, for instance, by belonging to a particular genre or featuring a particular performer, ensemble, or musical instrument.[11] These are reference works analogous to topical bibliographies, designed to scope out the universe of recordings relevant to some area of interest. Yet another type of discography functions as a works-cited list for an article or monograph and is intended only to identify the sources referenced in the body of the accompanying text.

So which of these models best suits our present needs? At minimum, we want to identify the specific audio sources referenced in the text, such as Berliner 638Y, to satisfy the basic expectations for a works-cited list. But for anyone who might want to investigate "Street Fakir" further, it's arguably just as important to know about the multiform nature of the work itself as it is to know that Berliner 638Y was the specific source we consulted for it. It's true that this condition doesn't uniformly affect all early recordings. Some of the selections discussed here are known to exist in just one single take, including George Graham's "Advertising Plant's Baking Powder" and "Free Silver Orator." But whenever multiple takes do exist, we believe it's worthwhile to draw this fact to the attention of any readers who care enough to look up a citation. And quite apart from the specifics of *these* selections, this book is likely to give at least some readers their first exposure to *any* early commercial spoken-word recordings and to influence how they think about them as objects of study in general. From that standpoint, too, it seems useful for us to list sources in a way that will give readers a sense of what it's like in practice to work with and seek out this kind of material. For all these reasons, we've tried to list, for each selection, not just the specific version cited in the text but all known versions for which someone might reasonably search. In each case, the entry for whichever source was consulted and cited in the text is marked with an asterisk; if an entry refers to multiple takes, the cited take is underlined. We considered including unissued masters, such as Victor M-2617-1, a take of George Graham's "Street Fakir" that is listed in the Victor ledgers but that was, as far as we know, never put into commercial production. In the end, we opted not to include these, in part because the recordings themselves are so unlikely to turn up and in part because our knowledge about unissued takes is inconsistent from company to company. (Besides, most available information about unissued takes could be readily obtained by searching DAHR.) Another more ambitious alternative would have been for us to compile a topical discography of all selections relevant to the concerns of this book—that is, of all known phonographic representations of storytelling, oratory, and so on. But that didn't seem practicable. A thorough discography of Cal Stewart's "Uncle Josh" routines alone would run to many pages. Hence, our discography is limited to those specific selections referenced elsewhere in the book (excluding this present essay). Finally, there's the matter of chronological and geographic scope. Recordings have continued to be made into the twenty-first century of some of the selections discussed here, such as Lincoln's Gettysburg Address, but listing all of them up to the present would seem unproductive. The latest recording cited in the text dates from 1922, so we've taken that year as our cut-off date, excluding—for example—Louis Armstrong's 1938 recordings of the two Elder Eatmore sermons.[12] Geographically, we've limited coverage to the United States, disregarding—among other things—a reported issue of "Jim Jackson's Race Track Story" on the Nipponophone label in Japan.[13] I've also noted any CD reissues of which I'm aware, regardless of their country of origin.

The titles associated with phonographic selections weren't standardized and can vary from version to version or even from pressing to pressing, as well as between ledgers, labels, catalogs, and spoken announcements. We've already seen this with "Street Fakir" and its variant title "Imitation of a Street Fakir," but the variation often takes forms that could pose a greater obstacle to searching or recognition. For example, recordings of George Graham's "Colored Funeral" routine can also be found labeled "Negro Funeral" and "Negro Funeral Sermon," but searching a resource such as DAHR for the title "Colored Funeral" wouldn't turn up versions with the title "Negro Funeral," and vice versa. In our discography, we've chosen one representative title for each selection and indicated any significantly different alternate titles next to the pertinent entries. It's worth noting that titles alone aren't a reliable basis for judging whether two recordings are versions of the "same" selection. So, for example, "The Old Time Street Fakir" and "The Patent Medicine Man" might look like plausible alternate titles for "Street Fakir," but listening to recordings with these titles reveals that they were applied to entirely different routines that don't share any overlap with Graham's, except for the fact that they also depict pitches for patent medicines.[14] In some cases, the content itself can leave room for uncertainty as to whether we're dealing with two variants of the "same" selection or two different selections. Consider "Address of the Late President McKinley at the Pan American Exposition," also issued as "Portion(s) of the Last Speech of President McKinley," "McKinley's Last Speech," and "President McKinley at the Pan American Exposition"—in each case, a recitation of part of the last speech McKinley delivered before his assassination, though not necessarily the same part. In addition to these recordings, there were also others that sandwiched a recitation of part of the same speech between musical performances by a band or vocal quartet, issued under titles such as "McKinley Memorial."[15] We haven't included the versions with music in our discographic listing because recording companies at the time offered the musical and non-musical versions separately and simultaneously, implying that they thought of them as different selections. But this was a judgment call that could have gone either way.

We've provided audio for the principal recordings discussed in this book on a companion website, so you shouldn't need to go hunting to listen to them, and the numbering of entries in the discography follows the sequence of those audio files. However, our discography also lists a number of other versions and takes, including some we haven't yet located ourselves, so some remarks are in order as to how you might go about trying to find and access these—or, for that matter, any other recordings from the same period you may find listed in DAHR or elsewhere. Standard bibliographic citations for books and journal articles dovetail neatly with the descriptive practices of libraries, publishers, and bookdealers, supporting well-established mechanisms of identification and retrieval. There still aren't any fully equivalent mechanisms for locating early commercial sound recordings. The current interlibrary loan system would be ill equipped to field a request for, say, "Jim Lawson's Hoss Trade with Deacon Witherspoon" by Cal Stewart on Victor M-1475-2. Even so, the conditions of access to this kind of material have improved dramatically in recent decades, and they're likely to improve yet further with the establishment of a legally recognized public domain in sound recordings in the United States beginning on January 1, 2022, as a provision of the Music Modernization Act of 2018. Twenty-five years ago, researchers who wanted to study early sound recordings in any serious way typically had either to arrange to visit a specialized library or archive in person or to immerse themselves in the world of private record collecting. Today, by contrast, a vast amount of this material is available digitally online. DAHR itself provides convenient links from discographic entries to whatever corresponding audio is available via the Library of Congress's National Jukebox, the Internet Archive's Great 78 Project, and the library of UCSB. Other important online repositories include the UCSB Cylinder Audio Archive (cylinders.library.ucsb.edu) and the Library of Historical Audio Recordings at i78s (i78s.org). There's even a significant amount of legacy audio available on YouTube, although its quality and documentation vary greatly (much of it consisting of video of records being played on vintage machines, uploaded by collectors who are more excited about the machines than the records). An ever-increasing amount of professionally restored content is also available for purchase from commercial reissue companies whose work deserves support. There's no longer any shortage of early material available for interested parties to hear with minimal effort. At the same time, though, it can still be challenging to find any one specific recording from the late nineteenth or early twentieth century. It's easy to sample what was recorded, but completeness can be a difficult goal, or even an impossible one.

Let's turn once again to "Street Fakir" for an illustration. We were originally exposed to this sketch by a well-worn original shellac pressing of Berliner 638Y I bought on eBay around 1997. It isn't particularly rare as Berliner discs go, and if I hadn't won that auction, I would have had other opportunities to try to pick up a copy in the meantime. Leafing through a stack of old auction catalogs from Nauck's Vintage Records, for example, I see that they offered another copy for sale in 2005.[16] But then I also happen to have a nice variable-speed turntable, styli of appropriate shapes and sizes, and a flat preamp, and I'm not afraid to use them. Most readers, I suspect, wouldn't have suitable equipment at hand for playing a record like this one and so might hesitate to bid on it, or they might have difficulty finding a way to access its content even if they obtained it, and a price well upward of one hundred dollars might be an added disincentive. Fortunately, audio for Berliner 638Y is also available online from the Library of Congress as a more universally serviceable MP3.[17] That was the only version of "Street Fakir" I'd heard before I began writing this essay, but when I looked the title up in DAHR and found listings for several other takes, I decided to try to locate copies of them in hopes that they could form the basis for a compelling case study in variability—and I wasn't disappointed. DAHR itself already provided a link to audio for Victor M-2617-2 at the UCSB Library, and Berliner 0585 and Zonophone 9304 became available soon afterward with the public launch of i78s in May 2021. Berliner 638 (with no take letter) was slightly more elusive, but a Google search led me to an MP3 at centuryoldsounds.com, the personal website of the collector Kevin D. Davis; in his accompanying text, Davis states that the take is "unknown," but he also provides an image of the disc's label area, which confirms that he has the unlettered take.[18] Thus, all the versions of George Graham's "Street Fakir" quoted in transcription earlier in this essay are readily accessible online as of this writing. However, one of the takes reportedly issued on disc is not: Victor V-2617-2, the seven-inch counterpart to M-2617-2. Does it contain the testimonial about President Roosevelt's voice and teeth—maybe in abridged form, to fit the lesser capacity of the seven-inch format? I don't know. The fellow collectors I've asked tell me they've never seen a copy, and I've found no trace of one in my searches of library catalogs and past auction records either. It's true that new content is constantly being added online, so perhaps audio for the missing take will eventually turn up on one of the sites I've mentioned or on some other site. On the other hand, if I decided this loose end was worth pursuing more proactively, I could post a query to relevant email discussion groups, such as 78-L or ARSCLIST, or to a bulletin board, such as the Talking Machine Forum. I could also keep an eye out for V-2617-2 at phonograph shows and in auction catalogs, and I could save an eBay search so that I'd be notified automatically of any new listings with the terms "George Graham" and "Street Fakir." These latter tactics often yield results, but a wait of several years wouldn't be unusual, and some items just seem never to surface at all. Meanwhile, there are other versions of "Street Fakir" besides Victor V-2617-2 that I haven't yet managed to find, and which could warrant similar searches: most notably, George Graham's brown wax cylinders, each of which would probably hold a unique recording, and a few takes on disc documented as having been recorded in England by other performers.[19] Any of these could provide an interesting surprise—most likely yet another testimonial involving some other public figure of the day. But until we track them down and hear them, we can only guess at what they might contain.

Throughout the period covered in this book, the pool of professional phonographic performers was small and specialized and didn't share much overlap with the big names of live concert or theatrical performance. Several interrelated factors were responsible for this situation. At first, the practical need to keep up inventories—particularly for the cylinder phonograph—meant that phonographic performers had to spend many grueling hours in the recording laboratory, often repeating the same selections over and over again to the limits of their endurance. Such work would have been hard to balance with a prominent stage career. But at the same time, only a few performers found themselves able to achieve consistently satisfactory results with the recording technology of the time, and that hurdle remained for a while even after large-scale duplication became feasible. Until the mid-1920s, commercial recording technology was acoustic rather than electric, which is to say that it relied wholly on the mechanical force of sound waves themselves to drive the process, with no help from artificial amplification. A spoken-word performer, for example, needed to talk straight into a funnel or tube, directing the sound waves of the voice against a diaphragm whose vibrations in response would then physically cause a stylus attached to the other side to move up and down, or from side to side, to produce modulations in a groove. Compared to the electric recordings that came later, these acoustic recordings suffered from a limited frequency range and from other sources of distortion, and by the same token, they were more forgiving or flattering to certain vocal qualities and techniques than to others. "There are not many people who make a success of it," one prominent phonographic vocalist remarked in 1905, "owing to the fact that it requires an iron throat, powerful lungs and a peculiar singing voice."[20] What was true of singing voices was also true of speaking voices, and such performers as Len Spencer, Cal Stewart, and Charles Ross Taggart evidently had whatever aptitude it took.

It wasn't just the recording side of the technology that was different. Playback at the time was acoustic as well, its sounds generated by direct mechanical linkage with the action of needles in record grooves with further coloring from attached horns. Today, chances are good that you'll experience these recordings played electrically rather than acoustically, which is as it should be: modern electric pickups are gentler on vulnerable analog records and superior in the accuracy of the signal they can retrieve. Chances are also good that the audio you hear will have undergone some subjective processing—and that's also as it should be. A "flat" transfer digitized straight from the output of an electromagnetic cartridge may be the best thing for preservation into the remote future, but with its intense high-frequency hiss and weak bass it's not optimal for actual listening. Instead, audition copies tend to have high frequencies reduced, low frequencies boosted, impulse noises (pops, clicks, crackle) suppressed, and perhaps some attenuation of broadband noise besides. Depending on skill and discretion, these processes can render a recording more intelligible and pleasant or riddle it with perceptually intrusive digital artifacts. The audio files presented on the companion website for this book have been only minimally processed; to experience what a virtuosic professional restoration sounds like, the reader is encouraged to search elsewhere.

INTRODUCTION

1. "Street Fakir"

George Graham
*Berliner 638, <u>638Y</u>, 7-inch disc, May 23, 1896, https://purl.dlib.indiana.edu/iudl /media/r46q97fk20
Berliner 0585, 7-inch disc, Oct. 11, 1899
Zonophone V9304, 7-inch disc, before Nov. 1900
Columbia 10504, 100 tpi cylinder, listed by June 1897, possibly 1896 **(as "The Street Fakir")**
(as "Imitation of a Street Fakir")
Victor V-2617-2, 7-inch disc, Apr. 22, 1903
Victor M-2617-2, 10-inch disc, Apr. 22, 1903

2. "Fakir Selling Corn Cure"

George Graham
*Berliner 639, 7-inch disc, May 23, 1896, https://purl.dlib.indiana.edu/iudl/media /g35445jz5m
Berliner 639Y, 7-inch disc, Oct. 15, 1897
Berliner 639U, 7-inch disc, Mar. 3, 1899

3. "Advertising Plant's Baking Powder"

George Graham
*Berliner 641, 7-inch disc, May 26, 1896; reissue: *Emile Berliner's Gramophone: The Earliest Discs, 1888-1901* (Symposium 1058), track 7, https://purl.dlib.indiana .edu/iudl/media/c77s65px6z

CHAPTER 1

4–5. "Lincoln's Speech at Gettysburg"

Anonymous (but often by Russell Hunting)
Columbia 3028, 100 tpi cylinder, listed 1895

Columbia 11004, 100 tpi cylinder, listed Nov. 1896
Russell Hunting
Edison 3821, 100 tpi cylinder, listed 1897 or 1898
William F. Hooley
*Berliner 6012, 7-inch disc, Sept. 21, 1898, https://purl.dlib.indiana.edu/iudl
/media/197x61mb69
Victor V-56-2, 7-inch disc, June 7, 1900
Victor V-56-5, 7-inch disc, Nov. 6, 1900
Harry Spencer
Zonophone X9233, 7-inch disc, before Oct. 1900
Columbia 160-1, 7-inch disc, listed 1902
Columbia 160-1,2, 10-inch disc, listed 1902
Columbia 160-8, 10-inch discs 160 and A280, in or before 1908
Leeds and Catlin matrix, 10-inch discs Imperial 44847 and Oxford 11847, ca. 1906
Len Spencer
Zonophone 1880, 7-inch disc, before May 1902
Zonophone 882, 9-inch disc, before May 1902
Edison 8154, 100 tpi cylinder, listed Aug. 1902
Victor M-2113-1, 10-inch disc, Mar. 21, 1903
Victor E-3939-1, 8-inch disc 2113, Oct. 26, 1906
*Victor B-3939-1, 10-inch discs 2113 and 16106-A, Oct. 26, 1906, https://purl.dlib
.indiana.edu/iudl/media/g94h640p74
Harry James (as "Lincoln's Gettysburg Address")
Brunswick A110, 10-inch disc 2770, May 15, 1924
Harry E. Humphrey
Columbia 38930-4, 10-inch disc A3044, July 17, 1913 **(as "Lincoln's Gettysburg Speech")**
Victor C-12837-2, 12-inch disc 35377-A, Feb. 20, 1914 **(as "Lincoln's Gettysburg
Address")**
Edison Blue Amberol 1651, 200 tpi cylinder, Mar. 10, 1915
Pathé E65559, 11½-inch disc B35074, ca. 1916

6. "Portions of the Last Speech of President McKinley"

Frank C. Stanley
Edison 7982, 100 tpi cylinder, listed ca. Oct. 1901 **(as "McKinley's Last Speech")**
William F. Hooley
Victor V-1070-1, 7-inch disc, Oct. 26, 1901
Victor M-1070-1, 10-inch disc, Oct. 26, 1901 **(as "Portion....")**
Leonard G. Spencer
Victor V-2170-1, 7-inch disc, Apr. 17, 1903
Victor M-2170-1, 10-inch disc, Apr. 17, 1903
*Victor B-3941-1, 10-inch disc 2170, Oct. 26, 1906,[21] https://purl.dlib.indiana.edu
/iudl/media/m31168f386

Harry Spencer
(as "Address of the Late President McKinley at the Pan American Exposition")
Columbia 31666, 100 tpi cylinder, listed Jan. 1902
Columbia 833-5, 7-inch disc, listed 1902
Columbia 833-10, 10-inch discs 833 and A278, listed 1902
(as "President McKinley at the Pan American Exposition")
Leeds and Catlin matrix, 10-inch discs Imperial 44849 and Oxford 11849, ca. 1906

7. *"Congressman Filkin's Homecoming"*

Byron G. Harlan, Steve Porter, and company
(credited to Byron G. Harlan)
*Columbia 19236-2, 10-inch disc A1036, Feb. 10, 1911, https://purl.dlib.indiana.edu
   /iudl/media/n79h14hf1n
(credited to Byron G. Harlan and Steve Porter)
Zonophone 10-inch disc 5722-A, Feb. 2, 1911
Victor 10006-1, 10-inch disc 16866-B, Feb. 27, 1911
Rex 506, 10-inch disc 5032-B, ca. 1914

8. *"Free Silver Orator"*

George Graham and Band
*Berliner 660, 7-inch disc, Nov. 30, 1896, https://purl.dlib.indiana.edu/iudl
   /media/9504952734

9. *"Republican Responsibility and Performance;*
Democratic Responsibility and Failure"

William Howard Taft
*Columbia 14503, 10-inch disc (coupled with 14505), Aug. 27, 1908; reissue: *In Their
   Own Voices: The U.S. Presidential Elections of 1908 and 1912*, 2 CDs, Marston 52028-2
   (2000), disc 1, track 24, https://purl.dlib.indiana.edu/iudl/media/d375844c35

10. *"Foreign Missions"*

William Howard Taft
*Edison 100 tpi cylinder 9996, listed July 1908; reissue: *Debate '08: Taft and Bryan
   Campaign on the Edison Phonograph*, Archeophone ARCH 1008 (2008), track 3,
   https://purl.dlib.indiana.edu/iudl/media/x31q47zg9p
Victor B-6637-1, 10-inch disc 16143-A, Aug. 5, 1908; reissue: *In Their Own Voices: The
   U. S. Presidential Elections of 1908 and 1912*, 2 CDs, Marston 52028-2 (2000), disc
   1, track 21
Columbia 14508, 10-inch disc (coupled with 14509; coupling later issued as A1013),
   Aug. 27, 1908
Columbia 40554, 100 tpi cylinder, Aug. 27, 1908

11. *"Why the Trusts and Bosses Oppose the Progressive Party"*

Theodore Roosevelt

*Victor C-12409-1, 12-inch disc 35250-A, Sept. 22, 1912; reissue: *In Their Own Voices: The U. S. Presidential Elections of 1908 and 1912*, 2 CDs, Marston 52028-2 (2000), disc 2, track 14, https://purl.dlib.indiana.edu/iudl/media/465544j32k

12. *"The 'Abyssinian Treatment' of Standard Oil"*

Theodore Roosevelt

*Victor C-12408-1, 12-inch disc 35249-B, Sept. 22, 1912; reissue: *In Their Own Voices: The U. S. Presidential Elections of 1908 and 1912*, 2 CDs, Marston 52028-2 (2000), disc 2, track 13, https://purl.dlib.indiana.edu/iudl/media/b98m414b3t

CHAPTER 2

13. *"A Revival Meeting at Pumpkin Center"*

Cal Stewart

*U-S Everlasting 1349-1, 200 tpi cylinder, listed ca. late 1911, https://purl.dlib
.indiana.edu/iudl/media/g05f36bs93
Edison Amberol 657, 200 tpi cylinder, listed Apr. 1911; also issued as Edison Blue
Amberol 2009

14. *"Colored Funeral"*

George Graham

*Victor M-982-1,2, 10-inch disc, Oct. 9, 1901; also identified as M-1862-1,2,[22]
https://purl.dlib.indiana.edu/iudl/media/b98m414b2h
Victor V-1862-5, 7-inch disc, Mar. 7, 1903
(as "Negro Funeral Sermon")
Berliner 689, 7-inch disc, June 22, 1897
(as "Negro Funeral")
Berliner 0587, take 2, 7-inch disc, Oct. 19, 1899
Berliner 0587, take 3, 7-inch disc, Nov. 15, 1899

15. *"The New Parson at the Darktown Church"*

Peerless Quartet
(as Columbia Quartette)
Columbia 85116, 6-inch 100 tpi cylinder, listed May 1907
Columbia 3629-1,2,3, 10-inch discs 3629 and A490, listed June 1907
(as anonymous "Quartette")
Albany Indestructible 649, 100 tpi cylinder, listed ca. Nov. 1907

(as "New Parson at the Darktown Church")
Zonophone mx. 7646, 12-inch discs 7032 (single-faced) and 4044 (double-faced), before June 1907
(as "New Parson at Darktown Church")
*Victor B-5081-1,3, 10-inch discs 5402 and 16186-A, Feb. 14, 1908, https://purl.dlib .indiana.edu/iudl/media/ro74855b1q

16. "Brother Jones' Sermon"

Ralph Bingham
*Victor B-21405-3, 10-inch disc 18587-B, Jan. 4, 1918, https://purl.dlib.indiana.edu /iudl/media/f16co8qb9x

17. "Elder Eatmore's Sermon on Throwing Stones"

Bert Williams
*Columbia 49644-3, 12-inch disc A6141, June 27, 1919; reissue: *Bert Williams: His Final Releases, 1919–1922*, Archeophone ARCH 5002 (2001), track 6, https://purl .dlib.indiana.edu/iudl/media/t44p58w953

18. "Elder Eatmore's Sermon on Generosity"

Bert Williams
*Columbia 49643-2,3, 12-inch disc A6141, June 27, 1919; reissue: *Bert Williams: His Final Releases, 1919–1922*, Archeophone ARCH 5002 (2001), track 5, https://purl .dlib.indiana.edu/iudl/media/c38653dv52

CHAPTER 3

19. "Jim Lawson's Horse Trade"

Cal Stewart
Victor M-3102-1, 10-inch disc, Feb. 9, 1901
Zonophone 9897, 7-inch disc, possibly issued under same number as 9-inch disc, before May 1901
Columbia 31574, 100 tpi cylinder, listed Aug. 1901
Zonophone 5495, 7-inch disc, before June 1903
Zonophone 5495, 9-inch disc, before 1904 (also associated with mx. 1609, ca. 1904)
Albany Indestructible 906 (mx. 409), 100 tpi cylinder, ca. Aug. 1908; reissue: *Cal Stewart: The Indestructible Uncle Josh*, Archeophone ARCH 5009 (2013), track 14
Edison 10070, 100 tpi cylinder, listed Feb. 1909
U-S Everlasting 1581, 200 tpi cylinder, listed Oct. 1912
(as "Jim Lawson's Horse Trade with Deacon Witherspoon")
Edison 7847, 100 tpi cylinder, listed June 1901

(as "Uncle Josh on Jim Lawson's Horse Trade")
Columbia 3021-1, 7-inch disc, listed ca. Feb. 1905
Columbia 3021-1,2, 10-inch discs 3021, listed Feb. 1905, and (take 1) A404
(as "Jim Lawson's Hoss Trade with Deacon Witherspoon")
Victor V-1475-2, 7-inch disc, July 14, 1902
Victor M-1475-2, 10-inch disc, July 14, 1902
Victor V-1475-6, 7-inch disc, Apr. 27, 1903
*Victor M-1475-6, 10-inch disc, Apr. 27, 1903, https://purl.dlib.indiana.edu/iudl
    /media/534f661toq
(as "Jim Lawson's Hoss Trade")
Victor A-570-1, 7-inch disc 1475, Oct. 20, 1903
Victor E-570-1, 8-inch disc 5010, Dec. 13, 1906
Victor B-570-1, 10-inch disc 1475, Dec. 13, 1906
Andrew Keefe (as "Uncle Josh on Jim Lawson's Horse Trade")
Leeds and Catlin matrix 8361, 10-inch discs Imperial/Sun 44856 and Oxford 11857,
    ca. 1906

20. *"The Farmer and the Hogs"*

Edwin Whitney
*Victor B-8055-1, 10-inch disc 16489-B, June 15, 1909, https://purl.dlib.indiana.edu
    /iudl/media/920f85ts7z

21. *"Uncle Jim's Racetrack Story"*

Len Spencer
*Victor B-1246-1,4, 10-inch disc 2790, Apr. 21, 1904, https://purl.dlib.indiana.edu
    /iudl/media/g64t54s418
Len Spencer and Al S. Holt (as "Jim Jackson's Race Track Story"); *Holt provides
    imitations of dog barking, horse whinnying, etc., not present in Victor version*
American 10¾-inch disc 031141, circa 1905
International Record Company matrix, 10-inch discs Excelsior/International 2303,
    circa 1905
Zonophone 7-inch disc 5910, before May 1904
Zonophone 9-inch disc 5910, before May 1904 **(label misprint: "Rollhack" for
    "Race Track")**

CHAPTER 4

22. *"A Meeting of the Ananias Club"*

Cal Stewart
*Victor M-3103-1, 10-inch disc, Feb. 9, 1901, https://purl.dlib.indiana.edu/iudl
    /media/415p295m43

Edison 7846, 100 tpi cylinder, listed June 1901, also issued as molded cylinder in 1902

Columbia 31573, 100 tpi cylinder, listed Aug. 1901 **(as "A Meeting of the Ananias Club at Pumpkin Center")**

Zonophone 606, 9-inch disc, before Oct. 1901 **(as "Meeting of the Annanias Club")**

Victor V-1476-1, 7-inch disc, July 14, 1902

Victor M-1476-1, 10-inch disc, July 14, 1902

Lambert 850, 100 tpi cylinder, ca. 1903 **(as "Uncle Josh at the Ananias Club")**

Victor M-1476-5, 10-inch disc, Apr. 27, 1903

Victor A-569-1, 7-inch disc 1476, Oct. 20, 1903

Victor B-569-1, 10-inch disc 1476, Oct. 20, 1903

Victor E-569-1, 8-inch disc 1476, Jan. 24, 1907

Victor B-569-1 [*sic*, re-used take number], 10-inch disc 1476, Jan. 24, 1907

23. *"War Talk at Pun'kin Center"*

Cal Stewart

Edison 3833-A,B,C, 10-inch disc 50260-R, May 29, 1915; take C also issued as Edison Blue Amberol 2657, 200 tpi cylinder; reissue (take C): *The Great War: An American Musical Fantasy*, 2 CDs, Archeophone ARCH 2001 (2007), disc 1, track 4

*Victor B-16102-1, 10-inch disc 17820-A, June 14, 1915, https://purl.dlib.indiana.edu/iudl/media/544b693v59

(as "War Talk at Pumpkin Center")

Columbia 45738-3, 10-inch disc A1797, June 2, 1915

Columbia 78471-6, 10-inch disc A1797, July 12, 1919

24. *"Uncle Josh Buys a Victrola"*

Cal Stewart

*Victor B-23118-1, 10-inch disc 18793-A, Aug. 11, 1919, https://purl.dlib.indiana.edu/iudl/media/089227tc67

CHAPTER 5

25. *"The Old Country Fiddler on Astronomy"*

Charles Ross Taggart

*Victor B-16721-2, 10-inch disc 18148-B, June 22, 1916, https://purl.dlib.indiana.edu/iudl/media/d56z908r6n

26. *"Old Country Fiddler and the Book Agent"*

Charles Ross Taggart

*Victor B-16725-2, 10-inch disc 17931-B, Oct. 28, 1915, https://purl.dlib.indiana.edu/iudl/media/q47r66qz89

27. *"Old Country Fiddler on the School Board"*

Charles Ross Taggart
*Victor B-16722-1, 10-inch disc 17910-B, Oct. 27, 1915, https://purl.dlib.indiana.edu
   /iudl/media/494v53rx71

28. *"The Old Country Fiddler in New York"*

Charles Ross Taggart
*Victor B-15532-2, 10-inch disc 17700-A, Dec. 21, 1914, https://purl.dlib.indiana
   .edu/iudl/media/d17c780b38

29. *"Old Country Fiddler at the Telephone"*

Charles Ross Taggart
*Victor B-18003-1, 10-inch disc 18148-A, June 21, 1916, https://purl.dlib.indiana
   .edu/iudl/media/049g15j212
Edison 8598-C, 10-inch disc 51048, Sept. 18, 1922; also listed as Edison Blue
   Amberol, 200 tpi cylinder 4668

30. *"Uncle Zed Buys a Graphophone"*

Charles Ross Taggart
*Columbia 78206-1, 10-inch disc A2890, Dec. 16, 1918, https://purl.dlib.indiana
   .edu/iudl/media/x91366cx8x

31. *"A Country Fiddler at Home"*

Charles Ross Taggart
*Edison 8468-A,C, 10-inch disc 51001-R, May 26, 1922; take A also listed as Edison
   Blue Amberol 4620, 200 tpi cylinder, https://purl.dlib.indiana.edu/iudl
   /media/188158j593

32. *"Old Country Fiddler at the Dance"*

Charles Ross Taggart
*Victor C-18015-1, 12-inch disc 35632-B, June 23, 1916; reissue: *Before the Big Bang:
   Country Music Origins in the Acoustic Era, 1890-1926*, 6 CDs, Archeophone
   (forthcoming), disc 3, track 17, https://purl.dlib.indiana.edu/iudl/media
   /2773865m1s

33. *"The Pineville Band"*

Charles Ross Taggart
*Victor B-15600-3, 10-inch disc 17794-A, Apr. 12, 1915, https://purl.dlib.indiana
   .edu/iudl/media/r171887j14

34. "The Village Gossips"

Cal Stewart and Steve Porter
*Edison Blue Amberol 1594-1,2, 200 tpi cylinder, listed Aug. 1912, https://purl.dlib
.indiana.edu/iudl/media/x02138rf2z
Edison 3768-A,B,C, 10-inch disc 50249, May 10, 1915
Cal Stewart and Byron G. Harlan
Victor B-16069-2, 10-inch disc 17854-A, June 4, 1915

NOTES

1. The print volumes of EDVR are Fagan and Moran 1983; and Fagan and Moran 1986. I follow Fagan and Moran's lead in using the letters *V* and *M* to differentiate disc size for early Victor catalog numbers and, by extension, to masters of the "pre-matrix" era. DAHR instead uses pre-matrix A (for seven-inch) and pre-matrix B (for ten-inch) in brackets, but this strikes me as liable to cause confusion with the later *A* and *B* matrix number prefixes, while *V* and *M* have the advantage of not overlapping with the later prefixes.

2. "George Graham's Benefit," *Washington Post*, October 25, 1895, 3.

3. Harry O. Sooy, "Memoir of My Career at Victor Talking Machine Company," accessed May 25, 2021, http://www.davidsarnoff.org/sooyh-maintext1898.html.

4. A pressing with the original catalog number crossed out and the new catalog number 75, as well as a distinctive "MADE IN CANADA" patent stamp, may be seen at https://www.bac -lac.gc.ca/eng/discover/films-videos-sound-recordings/virtual-gramophone/Pages/Item .aspx?idNumber=1007605011.

5. For the initial listing, see https://archive.org/details/ColumbiaPhonograph1889-1896.

6. It's true that other numbers are sometimes visible on pressings, including a mysterious sequence prefixed by the letter *M* that was in use from 1904 through 1908, but these appear to be redundant from the standpoint of identifying takes today, whatever purpose they may originally have served.

7. Before this point, the company's disc labels had read "Improved Gram-o-phone Record" or "Improved Record."

8. During the period when announcements were being erased, Columbia identified stampers first by a letter after the matrix number (e.g., 266-6-G) and then by a third number (e.g., 266-6-28). On Victor discs, stampers were instead identified by one or more tiny superscript characters molded into the shellac to the left of the catalog number (e.g., J5402). However, content varies so rarely among stampers that these distinctions can generally be ignored for purposes of citation. They are of interest mainly to collectors who value early pressings and to researchers who use the highest attested stamper numbers to estimate quantities pressed.

9. Based on a comparison of U-S Everlasting 1581 (200 tpi) with Albany Indestructible 906, Edison 10070, and Columbia 31574 (all 100 tpi).

10. This applies to later Edison molded cylinders, but the significance of similar dots on the first Edison molded wax cylinders remains unclear as of this writing.

11. Representative examples of the types mentioned are, respectively, Laird 1996; Brooks 2004b, 67–89; Smart 1970; and Heier and Lotz 1994.

12. Louis Armstrong, "Elder Eatmore's Sermon on Throwing Stones," Decca matrix 64436; and "Elder Eatmore's Sermon on Generosity," Decca matrix 64437, both August 11, 1938; coupled on Decca twelve-inch discs 15043 and 29231 and later reissued variously on LP and CD.

13. This issue, which isn't confirmed, would have used the same matrix as American 031141; see Bryant and Sutton 2015, 75.

14. Harlan and Porter, "The Old Time Street Fakir," Columbia 19303-3, ten-inch disc A1036, Apr. 20, 1911; Len Spencer, "The Patent Medicine Man," Victor M-2065-1, March 6, 1903. The latter title had been advertised on cylinder since the mid-1890s with a description confirming consistency of content, including a reference to the "Keeley cure"; see *Catalogue of Musical and Talking Records*, United States Phonograph Company [n.d.], reproduced in *The Thomas A. Edison Papers Digital Edition*, edison.rutgers.edu, CA029B. "The Medicine Fakir," listed during the brown wax era as Columbia 11026, was probably a variant title for "The Patent Medicine Man."

15. [Columbia Band], "McKinley Memorial," Columbia ten-inch disc 639-1, begins with "The Star Spangled Banner" and concludes with "Lead, Kindly Light." American Quartet, "Portions of the Last Speech of President McKinley and His Favorite Hymns," Victor M-1071-1, October 26, 1901, begins with "Lead, Kindly Light" and concludes with "Nearer, My God, to Thee."

16. Nauck's Vintage Records, *Vintage Record Auction #37*, closed May 7, 2005, lot 4.

17. George Graham, "Street Fakir," Library of Congress, accessed June 24, 2021, https://www.loc.gov/item/99390202/.

18. Century Old Sounds, "Berliner Disc Records, 1889–1900," accessed June 24, 2021, http://www.centuryoldsounds.com/Berliner.html.

19. Tom Collins, "The Street Fakir," Berliner (UK) 1012, October 12, 1898; Burt Shepard, "The Street Fakir," Gramophone and Typewriter 1110; also issued as Canadian Berliner 511; and possibly also Burt Shepard, "Whitechapel Street Fakir," Berliner (UK) 1058 (matrix 835), Jan. 13, 1899. I can't confirm that these are all versions of the "same" routine as George Graham's, but the first two appear alongside other selection titles also associated with him, suggesting that Collins and Shepard were copying his repertoire.

20. "Collins and Harlan at Milwaukee," EPM, June 1905, 13.

21. Neither M-2170-1 nor B-3941-1 has a matrix number visible at six o'clock, which appears to have caused some misidentification of these takes with one another, e.g., with audio links at DAHR.

22. Our source is a disc in my collection with no take number visible in the shellac at nine o'clock, which would ordinarily indicate a first take. DAHR doesn't list take 1 as issued, but it does list an "unknown" ten-inch take of March 7, 1903, as issued, and our take could conceivably also be this unusual "unknown" take rather than take 1. Its wording differs substantially from that heard on take M-982-2, which was reportedly recorded on the same day as take one, and my pressing bears a "Grand Prize" label consistent with a later date.

REFERENCES

Adorno, Theodor. 1990a. "The Form of the Phonograph Record." Translated by Thomas Y. Levin. *October* 55:56–61.

———. 1990b. "The Curves of the Needle." Translated by Thomas Y. Levin. *October* 55:48–55.

Agacinski, Sylviane. 2001. "Stages of Democracy." In *Public Space and Democracy*, edited by Marcel Hénaff and Tracy B. Strong, 129–143. Minneapolis: University of Minnesota Press.

Agha, Asif. 2007a. "Recombinant Selves in Mass Mediated Spacetime." *Language and Communication* 27:320–335.

———. 2007b. *Language and Social Relations*. Cambridge: Cambridge University Press.

Anderson, David. 2005. "Down Memory Lane: Nostalgia for the Old South in Post–Civil War Plantation Reminiscences." *Journal of Southern History* 71:105–136.

Anon. 1814. *The Charters and General Laws of the Colony and Province of Massachusetts Bay*. Boston: T. B. Wait.

———. 1878. "The Speaking Phonograph." *Scientific American Supplement* 115:1828–1829.

Atwood, E. Bagby. 1953. *A Survey of Verb Forms in the Eastern United States*. Ann Arbor: University of Michigan Press.

Auslander, Philip. 1999. *Liveness*. New York: Routledge.

Baer, Hans A., and Merrill Singer. 1997. "Toward a Typology of Black Sectarianism as a Response to Racial Stratification." In *African-American Religion: Interpretive Essays in History and Culture*, edited by Timothy E. Fulop and Albert J. Raboteau, 257–276. New York: Routledge.

Bailey, Liberty Hyde. 1915. *The Country-Life Movement in the United States*. New York: Macmillan.

———. (1911) 1917. *Report of the Commission on Country Life*. New York: Sturgis & Walton.

Bakhtin, M. M. 1981. *The Dialogic Imagination: Four Essays by M. M. Bakhtin*, edited by Michael Holquist. Austin: University of Texas Press.

Bartlett, Irving H. 1961. *Wendell Phillips, Brahmin Radical*. Boston: Beacon.

Baughman, Ernest Warren. 1966. *Type and Motif-Index of the Folktales of England and North America*. The Hague: Mouton.

Bauman, Richard. 1972. "The La Have Island General Store: Sociability and Verbal Art in a Nova Scotia Community." *Journal of American Folklore* 85:330–343.

———. 1977. *Verbal Art as Performance*. Prospect Heights, IL: Waveland.

———. 1986. *Story, Performance, and Event: Contextual Studies of Oral Narrative*. Cambridge: Cambridge University Press.

———. 2004. *A World of Others' Words: Cross-Cultural Perspectives on Intertextuality*. Malden, MA: Blackwell.

———. 2010. "'It's Not a Telescope, It's a Telephone': Encounters with the Telephone on Early Commercial Sound Recordings." In *Ideologies and Media Discourse: Texts, Practices, Policies*, edited by Sally Johnson and Tommaso Milani, 252–273. London: Continuum.

———. 2012. "Performance." In *A Companion to Folklore*, edited by Regina Bendix and Galit Hasan-Rokem, 94–118. Malden, MA: Wiley-Blackwell.

———. 2016. "Projecting Presence: Aura and Oratory in William Jennings Bryan's Presidential Races." In *Scale: Discourses and Dimensions of Social Life*, edited by E. Summerson Carr and Michael Lempert, 25–51. Berkeley: University of California Press.

Bauman, Richard, and Charles L. Briggs. 1990. "Performance and Poetics as Critical Perspectives on Language and Social Life." *Annual Review of Anthropology* 19:59–88.

———. 2003. *Voices of Modernity: Language Ideologies and the Politics of Inequality*. Cambridge: Cambridge University Press.

Benjamin, Walter. 1955. "The Work of Art in the Age of Mechanical Reproduction." In *Walter Benjamin: Essays and Reflections*, edited by Hannah Arendt, 217–251. New York: Schocken Books. Originally published 1936.

Betts, George Herbert, and Otis Earle Hall. 1914. *Better Rural Schools*. Indianapolis: Bobbs-Merrill.

Blommaert, Jan. 2015. "Chronotopes, Scales, and Complexity in the Study of Language in Society." *Annual Review of Anthropology* 44:105–116.

Bogle, Donald. 2001. *Toms, Coons, Mulattoes, Mammies, and Bucks: An Interpretive History of Blacks in American Films*. 4th ed. New York: Bloomsbury Academic.

Bolter, Jay David. 2016. "Remediation." In *The International Encyclopedia of Communication Theory and Philosophy*, edited by Klaus Bruhn Jensen and Robert T. Craig. Malden, MA: John Wiley. https://doi.org/10.1002/9781118766804.wbiect207.

Bolter, Jay David, and Richard Grusin. 1999. *Remediation: Understanding New Media*. Cambridge, MA: MIT Press.

Bowers, William L. 1974. *The Country Life Movement in America, 1900–1920*. Port Washington, NY: Kennikat.

Boyce, Adam R. 2013. *The Man from Vermont: Charles Ross Taggart, the Old Country Fiddler*. Charleston, SC: History Press.

Briggs, Charles L. 2007a. "Anthropology, Interviewing, and Communicability in Contemporary Society." *Current Anthropology* 48:551–580.

———. 2007b. "The Gallup Poll, Democracy, and the *Vox Populi*: Ideologies of Interviewing and the Communicability of Modern Life." *Text and Talk* 27:681–704.

Brooks, Tim. 2004a. *Lost Sounds: Blacks and the Birth of the Recording Industry, 1890–1919*. Urbana: University of Illinois Press.

———. 2004b. "George W. Johnson: An Annotated Discography." *ARSC Journal* 35:67–89.

———. 2020. *The Blackface Minstrel Show in Mass Media: 20th Century Performances on Radio, Records, Film and Television*. Jefferson, NC: McFarland.

Brooks, Tim, and Brian Rust. 1999. *Columbia Master Book Discography*. 4 vols. Westport, CT: Greenwood.

Brown, Carolyn S. 1987. *The Tall Tale in American Folklore and Literature*. Knoxville: University of Tennessee Press.

Brown, Dona. 1995. *Inventing New England: Regional Tourism in the Nineteenth Century*. Washington, DC: Smithsonian Institution.

Bryan, Mark Evans. 2002. *"Magnificent Barbarism": The Rube and the Performance of the Rural of the American Performance Stage, 1875–1925*. PhD diss., Ohio State University.

———. 2013. "Yeoman and Barbarians: Popular Outland Caricature and American Identity." *Journal of Popular Culture* 46 (3): 463–480.

Bryant, William R., and Allan Sutton. 2015. *American Record Company, Hawthorne & Sheble, International Record Company: Histories and Discographies, 1904–1909*. Denver: Mainspring.

Camlot, Jason. 2019. *Phonopoetics: The Making of Early Literary Recordings*. Stanford, CA: Stanford University Press.

Camus, Raoul F. 2013. "Band." In *The Grove Dictionary of American Music*, 2nd ed., edited by Charles Hiroshi Garrett. Oxford: Oxford University Press. http://doi .org/10.1093/acref/9780195314281.001.0001.

Canning, Charlotte M. 2005. *The Most American Thing in America: Circuit Chautauqua as Performance*. Iowa City: University of Iowa Press.

Case, Victoria, and Robert Ormond Case. 1970. *We Called It Culture: The Story of Chautauqua*. Freeport, NY: Books for Libraries. Originally published 1948.

Casey, Edward S. 2000. *Remembering: A Phenomenological Study*. 2nd ed. Bloomington: Indiana University Press.

Chambers, John Whiteclay. 2000. *The Tyranny of Change: America in the Progressive Era, 1890–1920*. 2nd ed. New Brunswick, NJ: Rutgers University Press.

Chanan, Michael. 1995. *Repeated Takes: A Short History of Recording and Its Effects on Music*. New York: Verso.

Cockrell, Dale. 1998. "Nineteenth-Century Popular Music." In *Cambridge History of American Music*, edited by David Nicholls, 158–185. Cambridge: Cambridge University Press.

Cogswell, Robert G. 1984. *Jokes in Blackface: A Discographic Folklore Study*. 2 vols. PhD diss., Indiana University, Bloomington.

Conforti, Joseph A. 2001. *Imagining New England: Explorations of Regional Identity from the Pilgrims to the Mid-Twentieth Century*. Chapel Hill: University of North Carolina Press.

Cubberly, Ellwood P. 1912. *The Improvement of Rural Schools*. Boston: Houghton Mifflin.

Danbom, David B. 1979. *The Resisted Revolution: Urban America and the Industrialization of Agriculture, 1900–1930*. Ames: Iowa State University Press.

———. 1995. *Born in the Country: A History of Rural America*. Baltimore: Johns Hopkins University Press.

Danielson, Virginia. 1997. *"The Voice of Egypt": Umm Kulthum, Arabic Song, and Egyptian Society in the Twentieth Century*. Chicago: University of Chicago Press.

Dargan, Amanda, and Steven Zeitlin. 1983. "American Talkers: Expressive Styles and Occupational Choice." *Journal of American Folklore* 96:3–33.

Davis, Gerald L. 1985. *I Got the Word in Me and I Can Sing It, You Know: A Study of the Performed African-American Sermon*. Philadelphia: University of Pennsylvania Press.

"Democratic Bureau's Suit against the M. A. Winter Company." 1900. *Phonoscope* 4 (6): 7.

Dent, Alexander Sebastian. 2009. *River of Tears: Country Music, Memory, and Modernity in Brazil*. Durham, NC: Duke University Press.

Diner, Steven J. 1998. *A Very Different Age: Americans in the Progressive Era*. New York: Hill & Wang.

Dorson, Richard M. 1946. *Jonathan Draws the Long Bow*. New York: Russell & Russell.

———. 1959. *American Folklore*. Chicago: University of Chicago Press.

DuBois, W. E. B. 1924. *The Gift of Black Folk: The Negroes in the Making of America*. Boston: Stratford.

———. 2003. *The Souls of Black Folk*. New York: Modern Library. Originally published 1903.

Dumont, Frank. 1899. *The Witmark Amateur Minstrel Guide and Burnt Cork Encyclopedia*. New York: M. Witmark & Sons.

Edison, Thomas A. 1878. "The Phonograph and Its Future." *North American Review* 126:527–536.

Elleström, Lars, ed. 2010. *Media Borders, Multimodality and Intermediality*. New York: Palgrave Macmillan.

Fabrizio, Timothy C., and George F. Paul. 2002. *Antique Phonography Advertising: An Illustrated History*. Atglen, PA: Schiffer.

Fagan, Ted, and R. William Moran. 1983. *The Encyclopedic Discography of Victor Recordings, Pre-Matrix Series*. Westport, CT: Greenwood.

———. 1986. *The Encyclopedic Discography of Victor Recordings, Matrix Series 1 through 4999*. Westport, CT: Greenwood.

Feaster, Patrick. 2001. "Framing the Mechanical Voice: Generic Conventions of Early Phonograph Recording." *Folklore Forum* 32:57–102.

———. 2006. "The Man Who Made Millions Laugh." Unpublished manuscript.

———. 2007. "'The Following Record': Making Sense of Phonographic Performance 1887–1908." PhD diss., Indiana University, Bloomington.

———. 2008. "Presidential Politics Meet the Talking Machine." Program notes for *Debate '08: Taft and Bryan Campaign on the Edison Phonograph*. Champaign, IL: Archeophone Records 1008.

Feld, Steven. 2015. "Acoustemology." In *Keywords in Sound*, edited by David Novak and Matt Sakakeeny, 12–21. Durham, NC: Duke University Press.

Fisher, Lawrence E., and Raven McDavid Jr. 1973. "Aphaeresis in New England." *American Speech* 48:246–249.

Foght, Harold Waldstein. 1910. *The American Rural School: Its Characteristics, Its Future and Its Problems*. New York: Macmillan.

Forbes, Camille. 2008. *Introducing Bert Williams: Burnt Cork, Broadway, and the Story of America's First Black Star*. New York: Basic Civitas.

Fox, Aaron A. 2004. *Real Country: Music and Language in Working-Class Culture*. Durham, NC: Duke University Press.

Gaisberg, Fred. 1942. *The Music Goes Round and Round*. New York: Macmillan.

Gal, Susan, and Judith T. Irvine. 2019. *Signs of Difference: Language and Ideology in Social Life*. Cambridge: Cambridge University Press.

Gale Research Company. 1983. *Currier & Ives: A Catalogue Raissoné*. 2 vols. Detroit: Gale Research.

"Gallery of Talent Employed for Making Records." 1898. *Phonoscope* 2 (7): 12–15.

Gelatt, Roland. 1997. *The Fabulous Phonograph 1877–1977*. New York: Macmillan.

Gershon, Ilana. 2017. "Language and the Newness of Media." *Annual Review of Anthropology* 46:15–31.

Gershon, Ilana, and Joshua A. Bell. 2013. "Introduction: The Newness of New Media." *Culture, Theory and Critique* 54 (3): 259–264.

Gershon, Ilana, and Paul Manning. 2014. "Language and Media." In *The Cambridge Handbook of Linguistic Anthropology*, edited by N. J. Enfield, Paul Kockelman, and Jack Sidnell, 559–575. Cambridge: Cambridge University Press.

Gilman, Sander L. 1974. *The Parodic Sermon in European Perspective: Aspects of Liturgical Parody from the Middle Ages to the Twentieth Century*. Wiesbaden, Germany: Franz Steiner Verlag.

Gitelman, Lisa. 1999. *Scripts, Grooves, and Writing Machines: Representing Technology in the Edison Era*. Stanford, CA: Stanford University Press.

———. 2006. *Always Already New: Media, History, and the Data of Culture*. Cambridge, MA: MIT Press.

Glassie, Henry. 2006. *The Stars of Ballymenone*. Bloomington: Indiana University Press.

Goffman, Erving. 1981. *Forms of Talk*. Philadelphia: University of Pennsylvania Press.

———. 1983. "The Interaction Order." *American Sociological Review* 48:1–17.

Gracyk, Tim. 2000. *Popular American Recording Pioneers, 1895–1925*. New York: Haworth.

Green, Lisa. 2002. *African American English: A Linguistic Introduction*. Cambridge: Cambridge University Press.

Greer, Lois Goodwin. 1927. "The Man from Vermont." *The Vermonter* 32 (6): 83–87.

Gumperz, John J. 1982. *Discourse Strategies*. Cambridge: Cambridge University Press.

Habermas, Jürgen. 1989. *The Structural Transformation of the Public Sphere: An Inquiry into a Category of Bourgeois Society*. Cambridge, MA: MIT Press. Originally published 1962.

Hahn, Steven, and Jonathan Prude, eds. 1985. *The Countryside in the Age of Capitalist Transformation: Essays in the Social History of Rural America*. Chapel Hill: University of North Carolina Press.

Harkness, Nicholas. 2011. "Anthropology at the Phonosonic Nexus." *Anthropology News* 52 (1): 5.

Harrison, Blake. 2005. "Tourism, Farm Abandonment, and the 'Typical' Vermonter, 1882–1930." *Journal of Historical Geography* 31:478–495.

Harrison, Harry P. 1958. *Culture under Canvas: The Story of Tent Chautauqua*. New York: Hastings House.

Hazeltine, Mayo W. 1902. *Orations from Homer to William McKinley*. Vol. 24. New York: P. F. Collier and Son.

Hazen, Margaret Hindle, and Robert M. Hazen. 1987. *The Music Men: An Illustrated History of Brass Bands in America, 1800–1920*. Washington, DC: Smithsonian Institution.

Heier, Uli, and Rainer E. Lotz. 1994. *The Banjo on Record: A Bio-Discography*. Westport, CT: Greenwood.

Hénaff, Marcel, and Tracy B. Strong. 2001. "Introduction: The Conditions of Public Space; Vision, Speech, and Theatricality." In *Public Space and Democracy*, edited by Marcel Hénaff and Tracy B. Strong, 1–31. Minneapolis: University of Minnesota Press.

Hinson, Glenn. 2000. *Fire in My Bones: Transcendence and the Holy Spirit in African American Gospel*. Philadelphia: University of Pennsylvania Press.

Hofstadter, Richard. 1955. *The Age of Reform*. New York: Vintage.

Holmberg, Carl Bryan, and Gilbert D. Schneider. 1986. "Daniel Decatur Emmett's Stump Sermons: Genuine Afro-American Culture, Language and Rhetoric in the Negro Minstrel Show." *Journal of Popular Culture* 19 (4): 27–38.

Hrkach, Jack. 1998. "The Yankee (United States: Nineteenth-Century Theater)." In *Fools and Jesters in Literature, Art, and History: A Bio-Bibliographical Sourcebook,* edited by Vicki K. Janik, 500–507. Westport, CT: Greenwood.

Hughes, Muriel Joy. 1959. "A Word-List from Vermont." *Vermont History* 27:123–167.

Hurston, Zora Neale. 1934. *Jonah's Gourd Vine.* Philadelphia: J. B. Lippincott.

———. 2022. *You Don't Know Negroes and Other Essays.* Edited by Genevieve West and Henry Louis Gates. New York: HarperCollins.

Hutchby, Ian. 2000. "Technologies, Texts, and Affordances." *Sociology* 35:441–456.

Irvine, Judith T. 1979. "Formality and Informality in Communicative Events." *American Anthropologist* 81:773–790.

Jakle, John A. 1999. "America's Small Town / Big City Dialectic." *Journal of Cultural Geography* 18 (2): 1–27.

Jensen, Klaus Bruhn. 2016. "Intermediality." In *The International Encyclopedia of Communication Theory and Philosophy,* edited by Klaus Bruhn Jensen and Robert T. Craig. Malden, MA: John Wiley. Wiley Online Library. Accessed March 15, 2021. https://doi-org.proxyiub.uits.iu.edu/10.1002/9781118766804.wbiect170.

Johnson, Edward H. 1877. Letter to *Scientific American* 37:304.

Johnson, James Weldon. 1912. *The Autobiography of an Ex-Colored Man.* Boston: Sherman, French.

———, ed. 1931. *The Book of American Negro Poetry.* New York: Harcourt Brace.

———. 1976. *God's Trombones: Seven Negro Sermons in Verse.* New York: Penguin Books. Originally published 1927.

Johnson, Nan. 1993. "The Popularization of Nineteenth-Century Rhetoric: Elocution and the Private Learner." In *Oratorical Culture in Nineteenth-Century America: Transformations in the Theory and Practice of Rhetoric,* edited by Gregory Clark and S. Michael Halloran, 139–157. Carbondale: Southern Illinois University Press.

Johnstone, Barbara. 1990. *Stories, Community, and Place: Narratives from Middle America.* Bloomington: Indiana University Press.

———. 1996. *The Linguistic Individual: Self-Expression in Language and Linguistics.* Oxford: Oxford University Press.

Jones, Malcolm. 1997. "The Parodic Sermon in Medieval and Early Modern England." *Medium Aevum* 66 (1): 95–114.

Kallen, Horace M. (1924) 1998. *Culture and Democracy in the United States.* New Brunswick, NJ: Transaction.

Kenney, William Howland. 1999. *Recorded Music in American Life: The Phonograph and Popular Memory.* Oxford: Oxford University Press.

Kurath, Hans. 1939a. *Handbook of the Linguistic Geography of New England.* Providence, RI: Brown University Press.

———. 1939b. *Linguistic Atlas of New England.* 3 vols. Providence, RI: Brown University Press.

———. 1949. *A Word Geography of the Eastern United States.* Ann Arbor: University of Michigan Press.

Laird, Ross. 1996. *Moanin' Low: A Discography of Female Popular Vocal Recordings, 1920–1933.* Westport, CT: Greenwood.

LaRue, Cleophas J. 2000. *The Heart of Black Preaching.* Louisville, KY: Westminster John Knox Press.

Lempert, Michael, and Sabina Perrino. 2007. "Entextualization and the Ends of Temporality." *Language & Communication* 27:205–211.

Lempert, Michael, and Michael Silverstein. 2012. *Creatures of Politics.* Bloomington: Indiana University Press.

Lippman, Walter. 1922. *Public Opinion.* New York: Harcourt, Brace.

Locke, Alain, ed. 1992. *The New Negro.* New York: Simon & Schuster. Originally published 1925.

Mahar, William J. 1985. "Black English in Early Blackface Minstrelsy: A New Interpretation of the Sources of Minstrel Show Dialect." *American Quarterly* 37 (2): 260–285.

———. 1999. *Behind the Burnt Cork Mask: Early Blackface Minstrelsy and Antebellum American Popular Culture.* Urbana: University of Illinois Press.

Marble, Ed. 1893. *The Minstrel Show, Or, Burnt Cork Comicalities: A Collection of Comic Songs, Stump Speeches, Monologues, Interludes, and Afterpieces for Minstrel Entertainments.* S.l.: s.n.

Marler, Scott P. 2003. "Country Store." In *Dictionary of American History,* edited by Stanley I. Kutler, vol. 2, 434–436. New York: Charles Scribner's Sons.

Martin, Lerone A. 2014. *Preaching on Wax: The Phonograph and the Shaping of Modern African American Religion.* New York: NYU Press.

Marvin, Carolyn. 1988. *When Old Technologies Were New: Thinking about Electric Communication in the Late Nineteenth Century.* Oxford: Oxford University Press.

McCurdy, Frances Lea. 1969. *Stump, Bar, and Pulpit: Speechmaking on the Missouri Frontier.* Columbia: University of Missouri Press.

McNutt, Randy. 1981. *Cal Stewart, Your Uncle Josh.* Fairfield, OH: Weathervane Books.

———. 2018. *Uncle Josh and the Hoosiers: Country Music's Pioneer.* Hamilton, OH: HHP Books.

Millard, Andre. 1995. *America on Record: A History of Recorded Sound.* Cambridge: Cambridge University Press.

Mitchell, Henry H. 1970. *Black Preaching.* Philadelphia: J. B. Lippincott.

Moore, Jerrold Northrop. 1999. *Sound Revolutions: A Biography of Fred Gaisberg, Founding Father of Commercial Sound Recording.* London: Sanctuary.

Moore, John Trotwood. 1906. "With Trotwood." *Trotwood's Monthly* 2 (1): 391–395.

Moore, Thomas. (1816) 1869. *Irish Melodies and Sacred Songs.* New York: Oakley, Mason.

Morgan, Winifred. 1988. *An American Icon: Brother Jonathan and American Identity.* Newark: University of Delaware Press.

Morton, David. 2000. *Off the Record: The Technology and Culture of Sound Recording in America.* New Brunswick, NJ: Rutgers University Press.

Nakassis, Constantine. 2016. *Doing Style: Youth and Mass Mediation in South India.* Chicago: University of Chicago Press.

Newsom, Jon. 1994. "The American Brass Band Movement in the Mid-Nineteenth Century." In *The Wind Ensemble and Its Repertoire: Essays on the Fortieth Anniversary of the Eastman Wind Ensemble,* edited by Frank J. Cipolla and Donald Hunsberger, 77–94. Rochester, NY: University of Rochester Press.

"New Use for the Talking Machine." 1900. *Phonoscope* 4 (4): 7.

Nickels, Cameron C. 1993. *New England Humor: From the Revolutionary War to the Civil War.* Knoxville: University of Tennessee Press.

"No Talking Machines for McKinley." 1900. *Phonoscope* 4 (6): 8.

Ochoa Gautier, Ana María. 2014. *Aurality: Listening and Knowledge in Nineteenth-Century Colombia.* Durham, NC: Duke University Press.

Oliver, Paul. 1984. *Songsters and Saints: Vocal Traditions on Race Records.* Cambridge: Cambridge University Press.

"Our Tattler." 1898a. *Phonoscope* 2 (8): 13.

"Our Tattler." 1898b. *Phonoscope* 2 (9): 9.

Petty, John A. 1974. "Cal Stewart: The Acoustic King of Comedy." *New Amberola Graphic* 11:1–7.

Pipes, William H. 1945. "Old-Time Negro Preaching: An Interpretive Study. *Quarterly Journal of Speech* 31:15–21.

———. (1951) 1992. *Say Amen, Brother: Old Time Negro Preaching; A Study in American Frustration.* Detroit: Wayne State University Press.

Pitts, Walter. 1989. "West African Poetics in the Black Preaching Style." *American Speech* 64 (2): 137–149.

Prescott, George B. 1877. "The Telephone and the Telegraph." *Scribner's Monthly* 15:848–858.

Proper, David R. 1998. "A Joyful Noise: 'Sounding Brass and Tinkling Cymbal': The Late Nineteenth-Century New England Town Band." In *New England Music: The Public Sphere, 1600–1900,* edited by Peter Benes, 160–175. The Dublin Seminar for New England Folklife Annual Proceedings 1996. Boston: Boston University.

Raboteau, Albert J. 1995. *A Fire in the Bones: Reflections on African-American Religious History*. Boston: Beacon.

Ray, Angela. 2005. *The Lyceum and Public Culture in the Nineteenth-Century United States*. East Lansing: Michigan State University Press.

Reed, Teresa. 2001. "Elder Eatmore and Deacon Jones: Folk Religion as Humor in Black Secular Recordings, 1918–1961." *Popular Music and Society* 25 (3): 25–44.

Rice, Tom. 2015. "Listening." In *Keywords in Sound*, edited by David Novak and Matt Sakakeeny, 99–111. Durham, NC: Duke University Press.

Rieser, Andrew C. 2003. *The Chautauqua Moment: Protestants, Progressives, and the Culture of Modern Liberalism*. New York: Columbia University Press.

Rosenberg, Bruce. 1988. *Can These Bones Live? The Art of the American Folk Preacher*. Rev. ed. Urbana: University of Illinois Press.

Rourke, Constance. (1931) 1959. *American Humor: A Study of the National Character*. New York: Harcourt Brace Jovanovich.

Rowland, Mabel. 1923. *Bert Williams, Son of Laughter: A Symposium of Tribute to the Man and to His Work by His Friends and Associates with a Preface by David Belasco*. New York: English Crafters.

Rugh, Susan Sessions. 2001. *Our Common Country: Family Farming, Culture, and Community in the Nineteenth-Century Midwest*. Bloomington: Indiana University Press.

Russell, Ian. 1991. "'My Dear, Dear Friends': The Parodic Sermon in Oral Tradition." In *Spoken in Jest*, edited by Gillian Bennett, 237–256. Sheffield, UK: Sheffield Academic.

Samuels, David W., Loise Meintjes, Ana Maria Ochoa, and Thomas Porcello. 2010. "Soundscapes: Toward a Sounded Anthropology." *Annual Review of Anthropology* 39:329–345.

Schafer, R. Murray. 1969. *The New Soundscape: A Handbook for the Modern Music Teacher*. New York: Associate Music.

———. (1977) 1994. *The Soundscape: Our Sonic Environment and the Tuning of the World*. Rochester, VT: Destiny Books.

Schiffrin, Deborah. 1981. "Tense Variation in Narrative." *Language* 57:45–62.

Shattuck, Herbert A. 1900. "June Notes." *Phonogram* 2:58–62.

Silverstein, Michael. 1996. "Monoglot 'Standard' in America: Standardization and Metaphors of Linguistic Hegemony." In *The Matrix of Language*, edited by Donald Brenneis and Ronald H. S. Macauley, 284–306. Boulder, CO: Westview.

———. 2004. "'Cultural' Concepts and the Language-Culture Nexus." *Current Anthropology* 45:621–652.

Simmel, Georg. (1908) 2009. *Sociology: Inquiries into the Construction of Social Forms*. Translated and edited by Anthony J. Blasi, Anton K. Jacobs, and Mathew Kanjirathinkal. Boston: Brill.

Simond, Ike. 1974. *Old Slack's Reminiscences and Pocket History of the Colored Profession from 1865 to 1891*. Bowling Green, OH: Popular.

Smart, James R. 1970. *The Sousa Band: A Discography*. Washington, DC: Library of Congress.

Smith, Eric Ledell. 1992. *Bert Williams: A Biography of the Pioneer Black Comedian*. Jefferson, NC: McFarland.

Smith, Jacob. 2008. *Vocal Tracks: Performance and Sound Media*. Berkeley: University of California Press.

———. 2015. *Eco-Sonic Media*. Berkeley: University of California Press.

Sterne, Jonathan. 2003. *The Audible Past: Cultural Origins of Sound Reproduction*. Durham, NC: Duke University Press.

Stewart, Cal. 1903. *Uncle Josh Weathersby's Punkin Centre Stories*. Chicago: Thompson and Thomas.

Stewart, James Brewer. 1986. *Wendell Phillips: Liberty's Hero*. Baton Rouge, LA: LSU Press.

Strausbaugh, John. 2006. *Black like You: Blackface, Whiteface, Insult and Imitation in American Popular Culture*. New York: Penguin Books.

Suisman, David. 2009. *Selling Sounds: The Commercial Revolution in American Music*. Cambridge, MA: Harvard University Press.

Taft, William Howard. 2001. *Political Issues and Outlooks: Speeches Delivered Between August 1908 and February 1909*. Athens: Ohio University Press. "Talking Machines in Politics." 1900. *Phonoscope* 4 (2): 6.

Tandy, Jeanette. 1964. *Crackerbox Philosophers in American Humor and Satire*. Port Washington, NY: Kennikat. Originally published 1925.

Tapia, John E. 1997. *Circuit Chautauqua: From Rural Education to Popular Entertainment in Early Twentieth-Century America*. Jefferson, NC: McFarland.

Thomas, Gerald. 1977. *An Analysis of the Tall Tale Genre with Particular Reference to Philippe d'Alcripe's La Nouvelle Fabrique des Excellents Traits de Vérité*. St. John's, Canada: Department of Folklore, Memorial University of Newfoundland in association with the American Folklore Society.

Tönnies, Ferdinand. (1887) 2001. *Community and Civil Society*. Translated by Jose Harris and Margaret Hollis. Edited by Jose Harris. Cambridge: Cambridge University Press.

Trachtenberg, Alan. 2007. *The Incorporation of America: Culture and Society in the Gilded Age*. New York: Hill & Wang. Originally published 1982.

Urban, Greg. 2001. *Metaculture: How Culture Moves through the World*. Minneapolis: University of Minnesota Press.

US Department of Agriculture, Office of the Secretary. 1915. *Social and Labor Needs of Farm Women*. Report no. 103. Washington, DC: Government Printing Office.

Walsh, Jim. 1951a. "Favorite Pioneer Recording Artists: Cal Stewart I." *Hobbies*, January 1951, 20–22.

———. 1951b. "Favorite Pioneer Recording Artists: Cal Stewart II." *Hobbies*, February 1951, 20–25.

———. 1951c. "Favorite Pioneer Recording Artists: Cal Stewart III." *Hobbies*, March 1951, 19–23.

———. 1951d. "Favorite Pioneer Recording Artists: Cal Stewart IV." *Hobbies*, April 1951, 20–24.

———. 1971. "Favorite Pioneer Recording Artists: My Last Words Concerning the Controversial McKinley Record, Part II." *Hobbies*, November 1971, 37–38, 48, 50, 92.

Warner, Michael. 2002. *Publics and Counterpublics*. New York: Zone Books.

Washington, Booker T. 1909. *The Story of the Negro: The Rise of the Race from Slavery*. 2 vols. New York: Doubleday, Page.

Washington, Booker T., N. B. Ward, and Fannie Barrier Williams. 1900. *A New Negro for a New Century*. Chicago: American.

Weidman, Amanda. 2015. "Voice." In Keywords in Sound, edited by David Novak and Matt Sakakeeny, 232–245. Durham, NC: Duke University Press.

Welch, Walter L., and Leah Brodbeck Stenzel Burt. 1994. *From Tinfoil to Stereo: The Acoustic Years of the Recording Industry 1877–1929*. Gainesville: University Press of Florida.

Welsch, Roger. 1972. *Shingling the Fog and Other Plains Lies*. Chicago: Swallow.

Wentworth, Harold. 1944. *American Dialect Dictionary*. New York: Thomas Y. Crowell.

Wharry, Cheryl. 2003. "Amen and Hallelujah Preaching: Discourse Functions in African American Sermons." *Language in Society* 32:203–225.

Wilson, Michael. 2006. *Storytelling and Theatre: Contemporary Storytellers and Their Art*. London: Palgrave Macmillan.

Wood, Joseph S. 1997. *The New England Village*. Baltimore: Johns Hopkins University Press.

Zboray, Ronald J. 1993. *A Fictive People: Antebellum Economic Development and the American Reading Public*. New York: Oxford University Press.

NEWSPAPERS AND PERIODICALS

Edison Phonograph Monthly (EPM)
New York Times (NYT)
Phonogram
Phonoscope
St. Louis Evening Post
Washington Post

AUDIO CD

Debate '08: Taft and Bryan Campaign on the Edison Phonograph. 2008. Archeophone Records 1008. Cited as *Debate*.

In Their Own Voices: The U.S. Presidential Elections of 1908 and 1912. Marston CD52028-2, 2000. Cited as *Voices*.

INDEX

communication technologies: cultural
inability to understand, 145; and youth
culture, 144
consumerism: and citizenship, 26, 36–37; the
phonograph in culture of, 113–14
Corbett, James J., 173, 175
country communicability: in city
encounters, 138–41; concept of, 119; music
in, 157; school boards and, 134–38; social
relations in, 156–57; Stewart's *versus*
Taggart's depictions of, 156–57; Taggart's
embodiment of, 119–20, 150; temporality
of, 129
country talkers, 118. *See also* Stewart, Cal;
storytelling; Taggart, Charles Ross
Cubberly, Ellwood P., 136–37

DAHR. *See* Discography of American
Historical Recordings
Danbom, David, 135
Davis, Kevin D., 189
dialects: African American dialect, 42–43,
57, 60, 62, 73; Graham's use of, 45, 48–49,
50, 51; minstrel, 48, 54, 55; rural vernacular
speech, 75, 77–78, 105; Stewart's use of
Yankee dialect, 77–78; Taggart's use of
Yankee dialect, 125–26
dialogue, cohesion devices in, 84
discographies: conventions in current book's
discography, 187; features and types of, 185
Discography of American Historical
Recordings (DAHR), 174, 183, 185, 186, 187,
188, 189, 199n1
discourse registers, 125
documentation of recordings: citation
styles and difficulties, 169–70, 179,
180, 182–83, 184, 186, 187–88, 199n8;
conventions used in this book, 186;
cylinder formats, 180–81; dates on, 183;
discographies and, 185; identification of
producers, 183–84; master recordings and
reproducibility, 181–82; matrix numbers,
182, 183–84; and multiple versions
and takes, 170–73, 175, 176–79, 185;
organization and numbering of, 182–83,

199n6; performers and, 182; recordings'
titles and ambiguities, 186–87
Du Bois, W. E. B., 62, 71, 72

Edison, Thomas: demonstrations of the
recording process by, 163–64; and
"The Liver Story," 74, 96n1; and
presidential campaign speeches,
26, 29; speculation on uses of recording
technology, 9; on use of the phonograph
for commemoration, 13
Edison Phonograph Company: adoption
of celluloid by, 185; advertisements by,
26, 31–32, 102; cylinders by, 181, 184–85,
199n10; Taggart recordings by, 122; tone
tests by, 165
Edison Phonograph Monthly: appeal to
collecting of record series, 108; on Bryan's
"Imperialism," 32; campaign speeches
advertising in, 26, 27; on lasting value of
recorded voices, 37; model advertising
letter in, 30–31; and political reach of
recordings, 31; on a record-taking event,
164–65
Encyclopedic Discography of Victor Recordings
(EDVR), 174, 182, 199n1

"The Farmer and the Hogs" (Whitney):
co-present assembled audience evoked
by, 94–95; the gentleman *versus* the
rustic in, 83; mimetic mode of character
presentation in, 83–84, 86; as remediated
performance, 86, 93–94; representation
of storytelling in, 94; transcript of, 80–82,
97n7
Feaster, Patrick, 116n2
Feld, Steven, 161
fiddling, traditional, 118. *See also* Taggart,
Charles Ross
films in political campaigns, 23–24, 36
"The First Phonograph Debate in History,"
29

Gaisberg, Fred, 2–3, 6
Gates, J. M., 45

RICHARD BAUMAN is Distinguished Professor Emeritus of Anthropology, of Folklore, and of Communication and Culture at Indiana University. He is author most recently of *A World of Others' Words: Cross-Cultural Perspectives on Intertextuality* and (with Charles L. Briggs) *Voices of Modernity: Language Ideologies and the Politics of Inequality*. He is coauthor (with Patricia Sawin and Inta Gale Carpenter) of *Reflections on the Folklife Festival: An Ethnography of Participant Experience* (Indiana University Press, 1992).

PATRICK FEASTER is Cofounder and Lead Researcher at First Sounds Initiative and former Media Preservation Specialist for Indiana University's Media Digitization and Preservation Initiative. He is a specialist in the history, culture, and preservation of early sound media and a three-time Grammy nominee.

www.ingramcontent.com/pod-product-compliance
Lightning Source LLC
Chambersburg PA
CBHW020348270326
41926CB00007B/353